FORENSIC
APPLICATIONS
OF
MASS
SPECTROMETRY

MODERN MASS SPECTROMETRY

Series Editor

Thomas Cairns, Ph.D., D.Sc.
Department of Health and Human Services
U.S. Food and Drug Administration
Los Angeles, California

Editorial Board

FORENSIC APPLICATIONS OF MASS SPECTROMETRY

Edited by

Jehuda Yinon, Ph.D.

Department of Environmental Science and Energy Research
Weizman Institute of Science
Rehovot, Israel

CRC Press
Boca Raton   Ann Arbor   London   Tokyo

Library of Congress Cataloging-in-Publication Data

Forensic applications of mass spectrometry / editor, Jehuda Yinon.
   p. cm. -- (Modern mass spectrometry)
  Includes bibliographical references and index.
  ISBN 0-8493-8252-1
  1. Chemistry, Forensic--Technique. 2. Mass spectrometry--Forensic applications. I.
Yinon, Jehuda. II. Series.
HV8073.5.F66 1995
363.2'562--dc20   94-17630
                            CIP

To my granddaughter, Shira

SERIES PREFACE

Publication of the first volume of the new series entitled 'Modern Mass Spectrometry' represents a milestone for scientists intimately involved with the practice of mass spectrometry in all its various forms. Forthcoming monographs in this series will focus on various selected topics within the rapidly expanding realm of mass spectrometry. Individual volumes will provide in-depth reports on mainstream developments where there is an urgent need for a specific mass spectrometry treatise in an active and popular area.

While mass spectrometry as a field is quite well-served by several publications and a number of societies, the application of mass spectrometric techniques across the basic scientific disciplines has not yet been recognized by existing journals. The present distribution of research and application papers in the scientific literature is widespread. There is a multidisciplinary audience requiring access to concise reports illustrating the latest successful approaches to difficult analytical problems.

The distinguished members of the Editorial Advisory Board all agreed that a platform exists for a premier book series with high standards to cover comprehensively general aspects of developing mass spectrometry. Contributing authors to the series will provide concise reports together with a bibliography of publications of importance selecting worthy examples for inclusion. Due to the rapid and extensive growth of the literature in mass spectrometry, there is a need for such reports by authorities who critique the entire subject area. There is an increasing urgency to provide readers with timely, informative and cogent reviews stripped of outdated material.

I believe that the decision to publish the series 'Modern Mass Spectrometry' reflects the realization that increasing numbers of mass spectrometrists are applying nascent state of the art approaches to some fascinating problems. Our challenge is to develop the best forum in which to present these emerging issues to encourage and stimulate other scientists to comprehend and adopt similar strategies for other projects.

Lofty ideals aside, our immediate goal is to build a reputable series with scientific authority and credibility. To this end, we have assembled an excellent Editorial Advisory Board that mirrors the prerequisite multidisciplinary exposure required for success.

I am confident that the mass spectrometry community will be pleased with the 'new publication approach' and will welcome the opportunities it presents fo foster the development of interactions between the various scientific disciplines. No doubt there is a long and difficult road ahead of us to ensure the series grows into a position of leadership. But I am conviced that the hard work of our outstanding Editorial Advisory Board, and the enthusiasm of our authors and readers will achieve the degree of success for which we all seek.

Thomas Cairns
Series Editor

FOREWORD

In this monograph of the series 'Modern Mass Spectrometry', seven chapters on state-of-the-art applications of mass spectrometry in the rapidly expanding field of forensic analysis have been written by international experts in their respective fields. The confidence in the reliability of mass spectrometry as an analytical tool is well-recognized for the strength of the technique in the areas of reproducibility, repeatability, specificity, and limit of detection. In a court of law, however, it is required to provide proof in a criminal case that is 'beyond a reasonable doubt' or in a civil matter that 'the preponderance of evidence' supports the conclusions.

In no other field has such intense development of methods been undertaken in support of collection of criminal evidence. The work presented in this book represents a quantum leap forward from where the field stood just a few years ago. Monitoring employees and athletes for evidence of potential abuse of drugs has become standard procedure in many corporate companies and various sports including the Olympic Games. Forensic scientists have quickly graduated to using more sophisticated instruments like the triple stage quadrupole instruments as well as the popular ion trap to help solve difficult chemical issues. Furthermore, the availability of mass spectrometry to provide reliable evidence at trace levels has caused the relatively new emergence of hair analysis for detection of drugs of abuse both for short- and long-term users. The use of isotope dilution mass spectrometry to detect adulteration of regulated products such as honey and various fruit juices can be considered a high-tech application with direct consumer benefits. Mass spectrometry is playing a more important role in our daily lives, both privately as well as professionally.

The Editor of this monograph, Dr. Jehuda Yinon, has a well-established international reputation in the area of forensic applications of mass spectrometry including identification of explosives. Seven keynote reviews capturing the expanding role of mass spectrometry in forensic science have been presented covering a wide variety of applications: drugs of abuse in body fluids, hair analysis for drugs of abuse, sports testing protocols, accelerants in fire debris, pyrolysis mass spectrometry of evidential materials such as paints, adhesives, etc., identification and characterization of explosives, and finally the use of isotope dilution mass spectrometry for authentication of natural products.

The chapters in this monograph represent a global perspective of the evolution of mass spectrometry to the forensic sciences. We know of no other easily accessible source of comprehensive, authoritative information as is offered in this book. We hope that such a comprehensive compilation will prove to be of great value to a multidisciplinary audience requiring access to concise reports illustrating the latest successful approaches to forensic analysis involving mass spectrometry.

Thomas Cairns
Series Editor

INTRODUCTION

Mass spectrometry, especially as a gas chromatograph/mass spectrometer (GC/MS) combination, is one of the major analytical pillars of the forensic laboratory. It is capable of separation and identification of minor constituents in complex matrices at a minimal time. The use of coated capillary columns and the use of alternate ionization techniques, such as electron impact (EI) and chemical ionization (CI), together with multiple ion detection (MID), have made GC/MS the method of choice of the forensic chemist.

With the commercial availability of interfacing techniques such as thermospray (TS) and particle-beam (PB), liquid chromatography/mass spectrometry (LC/MS) has become complementary to GC/MS, especially for nonvolatile and thermally labile compounds.

Tandem mass spectrometry (MS/MS) can provide an additional dimension of structural information for pure compounds and can serve as a tool for detection and identification of trace components in complex mixtures. This technique has now found its place in many forensic laboratories.

Mass spectrometry can also be used to "fingerprint" various compounds by measuring their isotopic composition. In combination with pyrolysis, mass spectrometry can provide information on inorganic solids.

The purpose of this book is to gather in one volume forensic applications of modern mass spectrometry. The subjects of the various chapters were chosen mainly according to the contribution of mass spectrometry as a major analytical tool in solving various forensic problems.

Mass spectrometry, and in particular GC/MS, is being extensively used for the detection and analysis of drugs of abuse and their metabolites in body fluids. A particular chapter has been devoted to mass spectrometry in sports testing, which deals with drug testing of urine samples of athletes in major sporting events. Forensic detection of drugs in hair has become of increasing interest and is the subject of a special chapter. Mass spectrometry has become in many forensic laboratories part of the routine analytical procedure for the identification of post-explosion residues. It has also been used in various combinations as a "sniffer" for the detection of hidden explosives. GC/MS methods have been developed for the identification of accelerants in contaminated extracts of suspected fire debris. Pyrolysis-mass spectrometry has been used for the characterization of synthetic polymers of forensic interest such as paint, fibers, etc.

Abundance ratios of stable isotopes, as measured by mass spectrometry, have been used as "fingerprints" to characterize and authenticate various compounds of forensic interest.

Mass spectrometry has become a well-established analytical technique in a large variety of forensic applications and is now an indispensable tool in forensic laboratories.

Jehuda Yinon

THE EDITIOR

Jehuda Yinon, Ph.D., is a Senior Research Fellow at the Weizmann Institute of Science, Rehovot, Israel. He received his B.Sc. and M.Sc. from the Technion, Israel Institute of Technology, Haifa, Israel, and his Ph.D. from the Weizmann Institute of Science. He was a Research Associate (1971 to 1973) and a Senior Research Associate (1976 to 1977) at the Jet Propulsion Laboratory, Pasadena, California, and a Visiting Scientist at the National Institute of Environmental Health Sciences in North Carolina (1980 to 1981), at the EPA Environmental Monitoring Systems Laboratory in Las Vegas, Nevada (1988 to 1989) and at the University of Florida in Gainesville, Florida (1993 to 1994).

Dr. Yinon's main research interests are environmental and forensic applications of mass spectrometry and the ion chemistry of energetic materials.

Dr. Yinon is a member of the American Society for Mass Spectrometry, the Israel Chemical Society, and the Israel Society for Mass Spectrometry. He is a Regional Editor of *Forensic Science Review* and is on the Editorial Boards of the *Journal of Energetic Materials* and of *Modern Mass Spectrometry*.

CONTRIBUTORS

Werner A. Baumgartner
Psychemedics Corporation
Culver City, California
and
Radioimmunoassay Laboratory
Nuclear Medicine Service
West Los Angeles Veterans Administration
Medical Center
Los Angeles, California

Wolfgang Bertsch
Department of Chemistry
The University of Alabama
Tuscaloosa, Alabama

Jean Louis Brazier
Laboratoire d'Etudes Analytiques et Cinetiques
du Medicament - LEACM
Institute of Pharmaceutical and Biological
Sciences
Institute of Industrial Pharmacy - IPIL
Lyon, France

Chen Chih Cheng
Psychemedics Corporation
Culver City, California

Bongchul Chung
Doping Control Center
Korea Institute of Science and Technology
Seoul, Korea

John T. Cody
Wilford Hall Medical Center
Lackland AFB, Texas

Thomas D. Donahue
Psychemedics Corporation
Culver City, California

Dean D. Fetterolf
Federal Bureau of Investigation
Laboratory Division
Forensic Science Research and Training Center
Quantico, Virginia

Rodger L. Foltz
Northwest Toxicology, Inc.
Salt Lake City, Utah

Gene F. Hayes
Psychemedics Corporation
Culver City, California

Virginia A. Hill
Radioimmunoassay Laboratory
Nuclear Medicine Service
West Los Angeles Veterans Administration
Medical Center
Los Angeles, California

Gunther Holzer
School of Biology
Georgia Institute of Technology
Atlanta, Georgia

Dongseok Lho
Doping Control Center
Korea Institute of Science and Technology
Seoul, Korea

Thomas O. Munson
Eckenfelder, Inc.
Nashville, Tennessee

Jongsei Park
Doping Control Center
Korea Institute of Science and Technology
Seoul, Korea

Songja Park
Doping Control Center
Korea Institute of Science and Technology
Seoul, Korea

Henry Scholtz
Psychemedics Corporation
Culver City, California

TABLE OF CONTENTS

Chapter **1**

GC/MS ANALYSIS OF BODY FLUIDS FOR DRUGS OF ABUSE

John T. Cody and Rodger L. Foltz

CONTENTS

0-8493-8252-1/95/$0.00+$.50
© 1995 by CRC Press Inc.

I. INTRODUCTION

The application of combined gas chromatography and mass spectrometry (GC/MS) to the identification and quantification of drugs of abuse in physiological specimens has grown dramatically in the past 10 years. This is particularly true in the U.S., primarily due to the increase in workplace drug testing and the widely accepted policy that whenever a drug analysis has the potential of legal and/or disciplinary consequences, the presence of the drug must be confirmed by a GC/MS analysis.

The purpose of this review is to summarize significant recent contributions to GC/MS methods for analysis of drugs of abuse in body fluids. Earlier reviews covered the same subject through 1985[1,2] and through 1990.[3] We have concentrated on covering the scientific literature from 1985 through 1993. We have also elected to use a different format from that followed in the earlier reviews. Each major drug of abuse is discussed in a separate section. Within each section the major issues and problems associated with development of a GC/MS assay for the drug are identified, followed by discussions of: (1) internal standards for quantitative analysis; (2) methods for hydrolysis (if required) and extraction of the drug from body fluids; (3) derivatization of the drug for GC/MS analysis; and (4) GC/MS conditions employed and performance reported for published procedures. We hope that this format will prove particularly useful for the analyst faced with the need to develop a drug assay with the sensitivity, specificity, quantitative accuracy, and cost constraints required for a specific analytical task.

II. GENERAL CONSIDERATIONS

A. Extraction

With rare exceptions, drugs and their metabolites must be extracted from biological matrices before analysis by GC/MS. The complexity of an extraction procedure depends on a number of factors that include the nature of the matrix and the target(s) of the analysis. If the analysis is designed to detect only a single compound, the extraction can be quite elaborate to isolate the compound of interest to the exclusion of all other compounds. On the other hand, if the procedure is designed to determine what compounds are present in the sample, a nonselective method of extraction is generally chosen.

Drug metabolism involves degradation and, in many instances, conjugation prior to excretion. Conjugated drugs and metabolites complicate extraction because of differences in solubility and the need for hydrolysis. Hydrolysis can be accomplished by any of several techniques. In one method, acid hydrolysis, the sample is acidified and then heated to hydrolyze the drug-conjugate linkage; this is a common technique for opiates. Alkaline hydrolysis is effective for ester-linked glucuronide conjugates, such as the major urinary metabolite of Δ^9-tetrahydro cannabinol. Enzymatic hydrolysis holds several advantages over nonenzymatic techniques. Enzymes are specific; they also allow the analysis of compounds that are unstable in the harsh environment of acid hydrolysis (i.e., acetylmorphine). Enzymatic hydrolysis typically produces cleaner extracts than does chemical hydrolysis.[4] A disadvantage is that enzymatic hydrolysis requires substantially more time than does acid hydrolysis; moreover, if compounds are conjugated by both sulfate and glucuronate, enzyme systems capable of hydrolyzing both types of conjugates should be used. Another potential difficulty is the somewhat sensitive nature of enzymes: appropriate sample pH and temperature are critical for enzymatic activity. Since the effectiveness of hydrolysis cannot be determined until a conjugated control has been measured, typically at the terminal stage of the analytical process, considerable time is lost if an enzyme proves ineffective. Acid hydrolysis, on the other hand, is not selective and is far more robust. It breaks both glucuronide and sulfate linkages and the process is faster than even the shortest enzymatic method.

An issue of growing concern from the viewpoint of health, safety, and environmental impact is the use of hazardous solvents. Acquisition, handling, storage, use, and disposal of these materials are being increasingly regulated. In the U.S., the Environmental Protection Agency has listed hazardous compounds which it considers to be particularly problematical from an environmental standpoint. A number of these compounds have been used for the extraction of drugs from biological fluids; they include benzene, chloroform, and diethyl ether, to name a few. Regulation of these and other hazardous chemicals is having a significant impact on the ways in which laboratories conduct their business. As regulatory requirements become ever more restrictive and costly, modifications must be made to procedures if laboratories are to operate within these requirements. The nature of a solvent and its volume each play a major role in these considerations. Procedures which minimize the use of solvents will have more widespread utility in the future.

A number of procedures are available for extracting multiple drugs from a sample. Some are designed to extract a group of drugs such as opiates, benzodiazepines, cocaine, and its metabolites, etc. Several procedures of this sort are described in the appropriate sections of this chapter. Still others are designed to extract as many drugs as possible from a sample; these procedures are primarily used in situations where it is important to identify all the drugs present rather than to identify or confirm a specific drug or metabolite. Many of the drugs described later in this chapter are detected with these broad screening procedures. A number of these screening techniques have been applied to emergency, clinical, and postmortem toxicology testing.[5-8]

1. Liquid-Liquid Extractions

Liquid-liquid extractions have a long history of use in toxicology laboratories. Important trends that have occurred in the development of liquid-liquid extraction procedures include: (1) the replacement of toxic and highly flammable solvents such as benzene and diethyl ether with similar but less hazardous solvents, such as toluene and methyl t-butyl ether, and (2) reduction in the volume of solvent used. The latter trend is a result of the high sensitivity afforded by GC/MS and the increasing use of deuterium-labeled internal standards which effectively compensate for incomplete recoveries of analytes. Liquid-liquid extractions can be very quick and simple, or complex and time-consuming, depending largely on the degree of selectivity required. With optimum selection of pH and extraction solvent, liquid-liquid extractions can be very selective and efficient. From a practical standpoint, however, the fine degree of refinement possible with liquid-liquid extractions is seldom employed and solid-phase extractions are generally as effective. However, the cost of materials for liquid-liquid extractions is generally lower than for solid-phase extractions, an important factor when assays involve large numbers of samples.

2. Solid-Phase Extractions

Solid-phase extraction has undergone significant changes in the last several years. Although the methodology is not new, the quality and consistency of commercially available solid-phase extraction materials have dramatically improved. In addition, the variety of adsorbents available has increased substantially.

Solid-phase extractions using a variety of packing materials have been applied to biological samples.[9,10] Solid-phase extraction also lends itself to automation. A number of manufacturers offer robotic systems designed to extract drugs of interest from biological fluids. Because of the large volume of urine drug testing, most of the development activity has been directed toward automating the extraction of drugs from urine. Adaptation of these procedures to other biological matrices is gaining momentum, however.

Automated solid-phase extraction systems range from simple liquid handlers, which can prepare columns, add the sample, wash, and finally elute the compound(s) of interest, to those that can literally start with a biological sample and end with the extract in an autosampler vial. These systems can also track the sample throughout the process to give a complete audit trail that will satisfy even the strictest chain-of-custody requirements.

A review of techniques for solid-phase extraction of drugs from biological tissues including brain, liver, intestine, kidney, bone, and adipose tissue has been compiled by Scheurer and Moore.[11] Other published reviews cover many of the options available to laboratories for solid-phase extractions from urine.[9,10]

B. Derivatization

Derivatization serves several important functions in GC/MS analysis. It can dramatically affect the volatility of a compound, improve its chromatographic behavior, and enhance the uniqueness of a compound's mass spectrum.

Large derivatives yield molecular and fragment ions with higher masses. While they are often advantageous for analysis of smaller drug molecules, such as amphetamines, the larger derivatives may not be appropriate for drugs like carboxy-THC and morphine, depending on the instrument being used. For example, some benchtop mass spectrometers have an effective upper mass limit of approximately 650 Da. However, the 3,6-bis-(heptafluorobutyryl) derivative of morphine has a mass of 677 Da. Similarly, carboxy-THC derivatized with pentafluoropropionic anhydride (PFPA) and hexafluoroisopropanol (HFIP) or pentafluoropropanol (PFP) has M^+ ions at 640 and 622 Da, respectively, both of which are at the high end of the mass range of benchtop instruments. Although such derivatives can be effective, the intensities of ions in the mid-mass ranges tend to be more reproducible than the intensities of ions at the extremes of an instrument's mass range.

Use of the same derivatizing reagent for different drug assays has several advantages. From a practical standpoint, preparation and use of one derivatizing reagent saves significant time and effort in the laboratory with respect to preparation and validation of the new reagent, and the training of personnel. For these reasons, derivatizing reagents such as PFPA and HFIP (or PFP), or bis-(trimethylsilyl)trifluoroacetamide (BSTFA), are popular in many laboratories because they can be applied to a variety of different assays.

Several extensive reviews of derivatization procedures and techniques have been published.[12-15]

C. Chromatographic Separation

With the nearly universal adoption of fused-silica capillary gas chromatographic columns by toxicology laboratories, the problem of choosing a suitable column for a particular drug assay has become far easier. No longer is it necessary to choose between a bewildering array of stationary phases. With few exceptions, a good quality immobilized methylsilicone or phenylmethylsilicone capillary column is capable of providing satisfactory chromatographic separation and resolution. Furthermore, due to the limited pumping capacity of benchtop GC/MS systems, narrow-bore (0.20 to 0.32 mm i.d.) columns are usually preferred because of their relatively low carrier-gas flow rates. As capillary columns with chiral stationary phases are expensive and have relatively low upper temperature limits, they have seen very limited use.

The method of injecting samples into the gas chromatographic column remains an important consideration. Although cold on-column injection and use of the "dropping needle injector" have strong proponents and are clearly very useful for certain types of analyses, splitless injection is undoubtedly the method most often used in the analysis of biological extracts. This is because of the sensitivity achievable by splitless injection and the technique's compatibility with automated sample injection. However, split injections have several significant advantages over splitless injection and should be considered whenever it is not necessary to achieve the very best sensitivity. Because split injections introduce less sample into the GC column, capillary columns last longer and ion sources require less frequent cleaning. Furthermore, in our experience the dynamic range is generally greater, peak widths are narrower, and ion ratios are more stable when samples are injected split as opposed to splitless. These advantages can be realized with only a very modest loss in sensitivity, if a split ratio in the order of 1:15 is used.

Methods of sample injection are discussed at length in two recent publications.[16,17]

D. Methods of Ionization

Electron ionization (EI) is the primary method of ionization in most forensic toxicology laboratories. Chemical ionization (CI) is the only other method of ionization currently used to an appreciable extent for forensic drug analysis. The continued dominance of electron ionization is due to a number of factors: (1) most mass spectrometers in forensic toxicology laboratories are not equipped for chemical ionization, (2) EI mass spectra are relatively reproducible and generally contain a detailed fragmentation pattern that, when matched with a reference spectrum, can constitute a conclusive identification, and (3) most analytical toxicologists are more familiar with electron ionization analyses than analyses based on other methods of ionization.

A limitation of electron ionization is that the molecular ion peak is often weak or totally absent. In contrast, chemical ionization can be used to clearly identify a compound's molecular weight. Chemical ionization is also highly versatile, in that a wide variety of spectra can be generated from the same analyte by simply changing the reagent gas; therefore, the analyst can tailor the mass analysis method for a particular analytical task.[18] This capability is most often called upon for those quantitative analyses where it is necessary to achieve the best sensitivity. Chemical ionization often provides better sensitivity than electron ionization because it can be more selective, and there is generally less fragmentation so that most of the ion current is concentrated at 1 m/z value. For example, use of ammonia as the reagent gas for chemical ionization of basic drugs generally results in a mass spectrum consisting of a single abundant ion corresponding to the protonated molecule (MH^+), or an ammonia-adduct ion (MNH_4^+). Since nonbasic compounds are not ionized efficiently by ammonia chemical ionization, they are unlikely to interfere with measurement of the drug analyte. Electron-capture negative-ion chemical ionization is also a relatively selective method of ionization; analytes containing halogens tend to be ionized far more efficiently than compounds that do not contain halogens. A potential disadvantage of chemical ionization is that CI mass spectra are more sensitive to ion-source conditions than EI mass spectra, and therefore tend to be less reproducible. Nevertheless, good quantitative reproducibility can be achieved with chemical ionization if ion-source conditions are carefully controlled and internal standards are well chosen.

E. Mass Analysis

The question of whether forensic samples should be analyzed by full-scan data acquisition or by selected ion monitoring has been hotly debated. The debate really comes down to a few basic questions: how much spectrometric and chromatographic information is needed to provide a scientifically and legally defensible identification? Is a full-scan spectrum necessary? If selected ion monitoring is acceptable, how many ions must be monitored? Unfortunately, there are no simple answers. A good quality full-scan mass spectrum that matches a reference spectrum clearly constitutes a more definitive identification than ion current profiles of a few selected ions. However, the specificity of a GC/MS assay depends on many factors, including (1) the choice of internal standard, (2) the selectivity of the extraction procedure, (3) the efficiency of the gas chromatographic separation, (4) the method of ionization and the relative uniqueness of the analyte's mass spectrum or the ions monitored, and (5) the signal-to-noise ratio of the detected ions. A well-designed assay involving selected ion monitoring of three or more abundant and structurally diagnostic ions, combined with precise determinations of the analyte's retention time relative to a suitable internal standard, can constitute a reliable identification, particularly if a reference standard of the analyte is analyzed within the same batch. Even fewer ions can provide an acceptable identification if the assay employs a highly selective extraction procedure and/or a selective method of ionization, such as ammonia chemical ionization. Selected ion monitoring typically provides signal intensities that are 10- to 100-fold better than those with full-scan analysis performed on the same instrument, and therefore it is better suited for quantitative measurements. Selected ion monitoring is also less subject to interferences from co-eluting compounds than an assay employing full-scan recording. However, unlike full-scan spectral acquisition, a selected ion monitoring assay will generally not detect the presence of unsuspected drugs that may be of forensic significance.

III. ANALYSIS OF COMMONLY ABUSED DRUGS

The drugs discussed in this review are those most often observed within the U.S. at the present time. However, drug abuse patterns are constantly changing, and they often differ from one country to another or even among regions within a country. Consequently, the forensic toxicologist must be prepared to modify existing assays or develop new procedures to identify and quantify new drugs of abuse.

Because of the current interest in the U.S. in workplace drug testing, published methods for detecting and measuring drugs of abuse in urine have proliferated and that trend is reflected in the content of this review. There has also been increased interest in the analysis of hair for drugs of abuse, which is reflected in recent reviews of the subject.[19-23] Consequently, GC/MS methods for hair analysis are not discussed at length in this review.

A. Cannabinoids

Marijuana is the most widely used illicit drug in the U.S. Consequently, most laboratories performing workplace drug testing conduct more GC/MS assays for 11-nor-9-carboxy-Δ^9-tetrahydrocannabinol (carboxy-THC), the major urinary metabolite of Δ^9-tetrahydrocannabinol (THC), than any other analyte. Detection of this metabolite in a blood or urine sample is considered conclusive evidence that the donor of the sample has recently used marijuana, or has been exposed to marijuana smoke, or has received a dose of the drug dronabinol (Marinol) which corresponds to synthetic THC. Marijuana contains many other cannabinoid compounds which are metabolized and excreted along with the carboxy-THC. Identification of urinary metabolites of other cannabinoids has been used as a means of determining whether a person has used marijuana or ingested dronabinol.

Because the number of cannabinoid confirmation assays performed is so large, a major concern in many laboratories is to minimize costs without sacrificing accuracy. Sensitivity of the assay is also a concern, since current guidelines require that confirmatory tests performed on specimens from Federal employees must be capable of reliably measuring carboxy-THC concentrations in urine below 15 ng/ml.[24]

Very little THC is excreted in urine. However, forensic investigators often request analysis of THC in a blood sample to gain insight as to whether use of marijuana could have been a contributing factor in an accident or crime. Determination of 11-hydroxy-Δ^9-tetrahydrocannabinol (HO-THC), a psychoactive metabolite, and carboxy-THC in blood may also be requested since the relative blood concentrations of THC, HO-THC, and carboxy-THC may suggest a time frame within which the marijuana entered the body.[25,26] For these reasons, this review includes recently described GC/MS methods for detecting and measuring THC and the two metabolites in blood as well as determination of carboxy-THC in urine. Earlier reviews of this subject covered material published before 1985.[27,28]

1. Internal Standards

Carboxy-THC containing 3 deuterium atoms attached to the 5' carbon has been widely used as the internal standard for quantitative analysis of this metabolite by GC/MS; this isotopomer is available from several commercial sources in the U.S. (Radian, Austin, TX; Sigma Chemical Company, St. Louis, MO; Alltech Associates, Deerfield, IL; and Research Triangle Institute, Research Triangle Park, NC). However, laboratories that derivatize carboxy-THC by methylation should be aware that the EI mass spectrum of methylated carboxy-THC shows a minor fragment ion at m/z 316, which coincides with the major ion in the EI mass spectrum of the corresponding trideuterated carboxy-THC derivative. Consequently, ion ratios which involve the m/z 316 ion will vary depending on the concentration of carboxy-THC. This problem can be avoided if the carboxy-THC isotopomer contains more than three deuteriums. Two such compounds are commercially available, carboxy-THC-d_6 (ElSohly Laboratories, Oxford, MS) and carboxy-THC-d_9 (Radian, Austin, TX). The major ions in the EI mass spectra of the most commonly used derivatives of carboxy-THC-d_6 were shown to be free of contributions from ions due to unlabeled carboxy-THC, even when the concentration of the unlabeled metabolite was 20-fold higher than the concentration of the internal standard.[29]

For GC/MS analyses employing full-scan data acquisition, hexadeuterated Δ^8-carboxy-THC (ElSohly Laboratories, Oxford, MS) is a suitable internal standard since its derivatives can be chromatographically separated from the corresponding derivatives of Δ^9-carboxy-THC.[30,31] The carboxy-THC-d_9 may be a reasonable alternative internal standard, since a good quality, narrow-bore capillary column should permit near baseline-separation of the carboxy-THC-d_9 from the unlabeled metabolite.

Trideuterated THC (available from Radian, Sigma Chemical Company, and Research Triangle Institute) is the recommended internal standard for determination of THC in physiological specimens.

Noncannabinoid internal standards are sometimes used for analysis of THC and its metabolites. However, a reported failure to detect carboxy-THC present in a urine specimen containing the metabolite illustrates what can happen when the internal standard is not an isotope-labeled analog.[32] In that case, the urine specimen also contained a high concentration of ibuprofen, which interfered with derivatization of the carboxy-THC but did not interfere with derivatization of the internal standard (1-pyrenebutyric acid). If deuterated carboxy-THC had been used as the internal standard, the failure to detect it would have clearly indicated a problem with either the extraction or derivatization of the sample.

2. Hydrolysis and Extraction

THC is highly lipophilic, readily soluble in nonpolar organic solvents, and only slightly soluble in water. When hexane is used to extract THC from plasma, recoveries tend to be variable. The drug can be extracted more efficiently if acetonitrile or methanol is first added to the plasma; then THC can be extracted from the supernatant with either hexane or another nonpolar solvent[33] or with a solid-phase extraction column.[34,35] The resulting extracts will still contain many endogenous lipids which can give rise to a high biological background in the subsequent GC/MS analysis, unless further purification is performed or a very selective method of GC/MS analysis is used, such as negative ion chemical ionization of the trifluoroacetyl derivative of THC[33] or tandem mass spectrometry.[34] Back-extraction of the THC into Claisen's alkali is a way of separating THC from coextracted neutral lipids.[25,36,37]

Determination of THC in adipose tissue requires an even more extensive sample cleanup, such as one developed for measurement of the drug in 100-mg fat-biopsy samples obtained from 11 marijuana users before and after they had smoked cigarettes containing deuterium-labeled THC.[38] After extraction with hexane-isopropanol (3:2), the THC was adsorbed onto prewashed Lipidex 5000 gel and eluted with methanol-water 0.14 M acetic acid (51:9:4). Further purification was achieved by solid-phase extraction with a C-18 column and elution with hexane. The extraction efficiency for the overall procedure was approximately 80%.

The carboxy-THC metabolite is excreted in urine primarily as the ester-linked glucuronide conjugate, which is readily hydrolyzed by exposure to alkali. After addition of 100 μl of 6 M sodium hydroxide to a 2-ml urine sample, hydrolysis is complete within 15 min at 25°C, or within 5 min at 50°C.[39] The glucuronide conjugate of carboxy-THC, commercially available from Alltech Associates (Deerfield, IL), can be used to verify completeness of hydrolysis. Neutral and basic compounds in the urine can be removed during the hydrolysis step by adding hexane to the alkalinized urine and agitating the sample. After centrifugation, the hexane layer is discarded; the aqueous layer is acidified and extracted with hexane-ethyl acetate (7:1).[33] Several published GC/MS assays for carboxy-THC in urine have invoked variations of this liquid-liquid extraction procedure.[40,41] However, solid-phase extractions are rapidly gaining in popularity. Commercially available solid-phase systems used for extraction of carboxy-THC from urine include: Bond Elut Certify II columns (Varian, Harbor City, CA),[42,43] C-18 Sep-Pak columns (Waters Associates, Milford, MA),[44] C-18 PrepSep columns (Fisher Scientific, Pittsburgh, PA),[45] Prep 1 anion-exchange columns (currently available from Creative Technologies, Wilmington, DE),[46,47] Supelclean Drug-Pak T columns (Supelco, Bellefonte, PA),[48,49] and SPEC extraction disks (Ansys, Irvine, CA).[31] Extraction efficiencies range from 50 to 90%. Solid-phase extraction with the SPEC system is unique in that the carboxy-THC is derivatized while adsorbed to the extraction disk, and no organic solvents are involved.[31] A highly selective antibody-mediated extraction reportedly permits measurement of sub-ng/ml concentrations of carboxy-THC in urine.[50]

Following oral administration of THC, carboxy-THC in blood is present primarily as the glucuronide conjugate.[51] However, after marijuana is smoked or THC is administered intravenously, a majority of the carboxy-THC in plasma is in the free form.[35,52] As the 8β,11-dihydroxy metabolite of THC is excreted in urine in a conjugated form, enzymatic hydrolysis is required before it can be measured by GC/MS analysis. McBurney et al. have suggested that detection of this metabolite in urine at concentrations above 15 ng/ml indicates that marijuana was used less than 6 h before the urine specimen was voided.[53]

3. Derivatization

Methods for derivatizing carboxy-THC for GC/MS analysis have been reviewed recently.[54] The most frequently reported methods involve either alkylation or silylation of the phenolic and the carboxylic groups. Alkylation can be performed by treatment with either methyl iodide[25,42,46,55,56] or n-propyl iodide[45] under strongly basic conditions. Alkylation has also been achieved by a procedure that combines the extraction and derivatization in one step.[39] After alkaline hydrolysis of the urine, the sample was mixed with 0.2 M tetrahexylammonium hydroxide and 0.2 M iodomethane in toluene for 30 min at 25°C. The toluene layer was passed through a glass pipet containing prewashed SM-7 acrylic copolymer beads

9-Carboxy-THC Derivatives

TABLE 1.

Relative Retention Times and Major Ions in the EI Mass Spectra of Derivatives of Carboxy-THC

R'	R"	Relative Retention Time	m/z of the Most Abundant Ions
–CH$_3$	–CH$_3$	1.000	372 (M$^+$), 357, 313
–C$_3$H$_7$	–C$_3$H$_7$	1.098	428 (M$^+$), 413, 385, 341
–Si(CH$_3$)$_3$	–Si(CH$_3$)$_3$	1.005	488 (M$^+$), 473, 371
–Sit–BuMe$_2$	–Sit–BuMe$_2$	1.170	572 (M$^+$), 515, 413
–CH(CF$_3$)$_2$	–COCF$_2$CF$_3$	0.740	640 (M$^+$), 489, 477, 429
–CH$_2$CF$_2$CF$_3$	–COCF$_2$CF$_3$	0.803	622 (M$^+$), 607, 459, 445

(Bio-Rad Labs, Sydney, Australia); the eluent was then concentrated by evaporation and injected into the GC/MS. The reported recovery of carboxy-THC by this procedure was 97%. The SM-7 beads were used to remove nonvolatile salts which would otherwise cause rapid deterioration of the capillary chromatographic column.

Silylation of the metabolite has been achieved by heating with BSTFA[40,41,48] or N-methyl-N-(tert-butyldimethylsilyl)-trifluoroacetamide (MTBSTFA).[57] Derivatization of both functional groups can also be achieved in one step by heating the extract with pentafluoropropionic anhydride and a fluorinated alcohol such as hexafluoroisopropanol or pentafluoro-propanol.[53,58,59]

Table 1 summarizes the relative retention times of various derivatives of carboxy-THC and lists the most abundant ions in the electron ionization mass spectra of each derivative. The relative retention times were determined on a 12-m DB-5MS capillary column with an internal diameter of 0.2 mm and a film thickness of 0.3 microns (J & W Scientific, Folsom, CA), temperature-programmed from 125 to 320°C at 18°/min.

Each method of derivatizing carboxy-THC has advantages. For example, the fluorinated derivatives have higher molecular weights and shorter GC retention times than either the silyl or alkyl derivatives. They also have high electron affinities, which permit measurement of very low concentrations of the metabolite when negative ion chemical ionization is used. The choice of derivative can have a significant effect on the limit of quantitation achievable. For example, in an unpublished comparison of sensitivities performed in one of the authors' (RLF) laboratories using electron ionization, the best sensitivity was achieved with the dipropyl derivative, followed in decreasing order of sensitivity by the dimethyl derivative, the bis-(trimethylsilyl) derivative, and the hexafluoroisopropyl-pentafluoropropionate derivative. The methyl and propyl derivatives of carboxy-THC are the most stable, followed by the *t*-butyldimethylsilyl derivative. The trimethylsilyl derivative is relatively stable as long as it is maintained in an excess of the trimethylsilylating reagent.

4. GC/MS Conditions and Performance

Capillary columns coated with a phenylmethylsilicone stationary phase are used most frequently for the GC/MS analysis of derivatized cannabinoids. The specific choice of capillary column is not critical as long as the column is relatively free of active sites. Derivatized extracts are normally injected in the splitless mode with typical temperature programs of 150 to 300°C at 15 to 25°/min. The major challenge is to achieve sufficient sensitivity to permit accurate measurement and reproducible ion intensities at concentrations in the very low ng/ml range. Most recently published GC/EI-MS assays for carboxy-THC report limits of quantitation between 1 and 10 ng/ml from 4 to 10 ml of sample. This sensitivity is adequate for most forensic purposes. However, concentrations of THC in blood typically fall below 1 ng/ml within a few hours after the subject has smoked a single marijuana cigarette. Consequently, better sensitivity is desirable when blood samples are analyzed for THC and its metabolites. A limit of quantitation (LOQ) of 0.2 ng/ml in 2 ml of blood has been reported for a THC assay based on negative ion chemical ionization of the trifluoroacetyl derivative.[33,60] In the same assay, the bis-trifluoroacetyl derivative of HO-THC was measured with an LOQ of 0.5 ng/ml, and the carboxy-THC was determined as the methyl ester trifluoroacetyl derivative with an LOQ of 0.1 ng/ml.[33]

An even more sensitive assay was recently developed for measurement of THC and HO-THC in plasma.[34] By combining positive ion chemical ionization of the trimethylsilyl derivatives of THC and HO-THC with gas chromatography and tandem mass spectrometry, the limits of quantitation were lowered to 0.05 and 0.1 ng/ml, respectively. Figure 1 shows the selected reaction current from analysis of a plasma sample containing 89 pg/ml of THC and 141 pg/ml of HO-THC.

B. Cocaine and Metabolites

Problems associated with the determination of cocaine and its two major metabolites, benzoylecgonine and ecgonine methyl ester, are most often due to the hydrolytic instability of the compounds[61-66] or to their differences in solubility. Each of the compounds is susceptible to degrada-

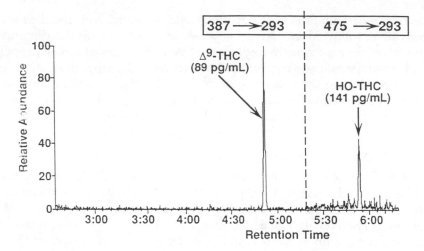

Figure 1.
The ion current profile from a GC/MS/MS analysis of a plasma sample containing
89 pg/ml of THC and 141 pg/ml of HO-THC.

tion under alkaline conditions; they are most stable at a pH of 5.[63] Blood
and plasma samples to be analyzed for cocaine should be stabilized with
sodium fluoride to prevent enzymatic degradation of the cocaine.[64,66,67]
Benzoylecgonine is a relatively polar amphoteric compound, whereas
cocaine is a basic lipophilic drug. Consequently, efficient extraction of
both compounds in a single-step procedure is difficult. Despite these
problems, determination of cocaine and its metabolites in biological samples
is a common task in most forensic toxicology laboratories due to the
widespread abuse of cocaine, particularly in the U.S. GC/MS methods for
analysis of cocaine and it metabolites were most recently reviewed by
Cone and Darwin.[68]

   Cocaine is derived from the *Erythroxylon coca* plant, which is grown in
several regions of the world and has a long and varied history. It is a
schedule II drug and its medical use is limited to relatively few situations.
Cocaine hydrochloride, the powdered form of the drug, is most often
taken by snorting and is readily absorbed into the blood stream through
the nasal mucosa. Inhalation of cocaine by smoking the free base form of
the drug, "crack", offers an even more rapid entry of the drug into the
bloodstream via the pulmonary circulation, resulting in a corresponding
increase in the speed of onset and intensity of the "high" perceived by the
user. Although the incidence of cocaine use relative to other drugs of
abuse has varied over the years, it is still widely abused and there is little
basis for hoping that its popularity will diminish substantially in the
foreseeable future.

   Once absorbed into the body, cocaine is rapidly metabolized to ecgonine
methyl ester, benzoylecgonine,[69–73] and a number of minor metabolites.[74,75]

Estimates of the half-life of cocaine in circulation range from 15 to 120 min. However, some chronic users of cocaine who stop use of the drug have been shown to continue to excrete cocaine in urine at concentrations exceeding 10 ng/ml for up to 15 days, presumably due to release of the drug from fatty tissue.[76] Approximately 80% of a dose of cocaine is excreted in the urine as the two major metabolites; benzoylecgonine is excreted more slowly than ecgonine methyl ester and can therefore be detected in urine for a longer period of time. As a result, detection of benzoylecgonine in urine is the most common analytical method of identifying cocaine users.

Cocaethylene, the ethyl ester of benzoylecgonine, was first shown by Rafla and Epstein in 1979 to be formed *in vivo* when cocaine and ethanol are consumed at the same time.[77] A growing interest in this metabolite has resulted in publication of GC/MS methods for its determination in plasma,[78] urine,[79] and hair.[68,80] Since cocaethylene is also formed *in vitro* from cocaine and ethanol,[81] artifactual formation of the compound is a possibility whenever cocaine is exposed to ethanol either before consumption or during an analytical procedure.

1. Internal Standards

Deuterium-labeled analogs currently available include: benzoylecgonine-d_3 (Alltech Associates, Deerfield, IL; Radian, Austin, TX; Sigma Chemical Company, St. Louis, MO), cocaine-d_3, ecgonine methyl ester-d_3, ecgonine-d_3 (Radian and Sigma Chemical Company), and cocaethylene-d_3 (Radian). As with most other analytes, the best quantitative results are obtained when the corresponding isotopomer is used as the internal standard for each analyte to be measured. However, several reported GC/MS assays for cocaine and its metabolites have used structurally related compounds rather than isotopomers as internal standards. For example, ketamine has been used as the internal standard for determination of cocaine, benzoylecgonine, and ecgonine methyl ester in urine.[82] Propylbenzoylecgonine served as the internal standard for the GC/MS analysis of cocaine and cocaethylene in blood and tissue samples.[78] However, this ester of benzoylecgonine is not a suitable internal standard for the analysis of benzoylecgonine due to the large difference in its solubility compared to that of benzoylecgonine. An exception to this statement is when the benzoylecgonine is derivatized during the extraction process, as in the extractive alkylation procedure reported by Joern.[83]

2. Extraction

Benzoylecgonine and ecgonine methyl ester are excreted in the urine as free compounds, but some of the minor metabolites are excreted as conjugates.[74] Cocaine is readily extracted from biological specimens at basic pHs by liquid-liquid or solid-phase procedures. However, cocaine is susceptible to hydrolysis at the high pH values sometimes encountered in extraction procedures, which results in conversion to benzoylecgonine.[63,84] Benzoylecgonine and ecgonine methyl ester are also unstable under alkaline conditions; benzoylecgonine is hydrolyzed to ecgonine,[85] and ecgonine methyl ester undergoes isomerization to pseudoecgonine methyl ester. Ecgonine methyl ester is relatively easily extracted and, although a derivatizable group is present on the molecule, it is often analyzed without derivatization. Benzoylecgonine, on the other hand, is more difficult to extract and requires derivatization before GC/MS analysis.

A reported liquid-liquid extraction of benzoylecgonine involved addition of 5 ml of ethanol:methylene chloride (1:1) to a 3-ml urine sample made basic by the addition of 0.2 ml of 3 M ammonium hydroxide; vortex mixing for 30 s was followed by centrifugation to separate the phases. This procedure yielded essentially 100% recovery of the metabolite from urine. Use of lesser amounts of ethanol gave inconsistent results, while a higher proportion of ethanol increased the tendency for emulsion formation.[86] Another liquid-liquid extraction procedure used chloroform:isopropanol (90:10) to extract benzoylecgonine, ecgonine methyl ester, and cocaine from urine with recoveries of 76, 72, and 82%, respectively.[82]

A surprisingly effective cleanup step performed before extraction of benzoylecgonine from urine was recently reported by Gerlits.[87] Urine samples were made basic by addition of 0.1 M phosphate buffer (pH 12) and washed with a solvent system consisting of equal proportions of methylene chloride, hexane, and diethyl ether. The benzoylecgonine was then extracted with methylene chloride with a recovery of approximately 68%, which compares favorably with previously reported liquid-liquid extraction procedures while providing interference-free extracts. Although the wash step was accomplished at a high pH, hydrolysis of cocaine in the samples was determined to be less than 0.5%. The dynamic range reported for this procedure was reported to be 75 to 1000 ng/ml.

A back-extraction was used to clean up extracts of blood and liver prior to GC/MS determination of cocaine and cocaethylene concentrations.[78] To 1 ml of blood or tissue homogenate was added 1 ml of saturated borate buffer (pH 8.9) and 10 ml of 1-chlorobutane. After mixing for 15 min, the 1-chlorobutane layer was back extracted with 1 ml of 0.1 M HCl. The acid layer was washed with 5 ml of hexane, alkalinized to pH 8.9, and extracted with 10 ml of 1-chlorobutane. The recoveries for cocaine and cocaethylene were 84 and 87%, respectively.

Cocaine and benzoylecgonine were extracted from blood and urine samples using Bond Elut C_{18} cartridges after addition of 2 ml of pH 9.0 buffer to a 1-ml sample. Blood samples were centrifuged and the fluid portion was poured into the extraction cartridge. This simple dilution and centrifugation proved effective even with very viscous postmortem blood samples. Recoveries were reported to be 86 and 75% for cocaine and 95 and 87% for benzoylecgonine from blood and urine, respectively.[88] Hemolyzation by sonication is an alternative method of preparing blood samples for solid-phase extraction.[89]

Bond Elut Certify cartridges were also used to extract reconstituted lyophilized human urine containing certified concentrations of cocaine and benzoylecgonine; recoveries of benzoylecgonine ranged from 76 to 96%.[90]

Benzoylecgonine was extracted from urine with Drug-Clean GC/MS SPE tubes by a procedure that does not require preconditioning of the extraction tubes.[91] The reported recoveries of the benzoylecgonine ranged from 74 to 100%.

Matsubara et al. extracted benzoylecgonine, cocaine, and ecgonine methyl ester using Extrelut columns, but the ecgonine methyl ester was extracted separately from the cocaine and benzoylecgonine.[92] With this dual extraction, the recoveries were 81, 95, and 97% for benzoylecgonine, cocaine, and ecgonine methyl ester, respectively.

An advantage of solid-phase extraction is that it is more easily automated than liquid-liquid extraction procedures. For example, Detectabuse extraction columns were used with a robotic system for extraction of cocaine and benzoylecgonine from urine.[93]

Many of the procedures developed for urine work well with other types of specimens; however, direct adaptation is not always effective. As an example, the manufacturer's recommended protocol for extraction of cocaine and benzoylecgonine from urine using Bond Elut Certify columns was found to be unacceptable for testing saliva, due to differing osmolarities.[94] Addition of 6 ml of an electrolyte solution to 1 ml of saliva proved an effective modification for isolating the drug from that unique matrix.

Concern over the use of illicit drugs during pregnancy has resulted in development of GC/MS methods for determination of cocaine and its metabolites in meconium.[95–97] Solid-phase extraction with Bond Elut Certify columns following an initial methanol extraction of meconium is described in two of the published procedures.[95,97] The extraction efficiencies for cocaine, benzoylecgonine, ecgonine methyl ester, and cocaethylene in meconium varied from 58 to 99% in one of the procedures; the extraction efficiencies were generally higher for blood samples.[95]

3. Derivatization

Cocaine contains no readily derivatizable functional groups; so GC/MS assays for only the parent drug do not include a derivatization step. Methods that include analysis for benzoylecgonine and/or ecgonine methyl ester normally do include a derivatization step. Derivatization of benzoylecgonine has been accomplished by either alkylation or silylation. Alkylation reactions have included formation of the methyl,[88] propyl,[98] pentafluoro-n-propyl,[82,95,97] or hexafluoroisopropyl[74,88] esters of benzoylecgonine. Methylation converts benzoylecgonine back to cocaine. However, if the methylation is performed with iodomethane-d_3 the product can be distinguished mass spectrometrically from any cocaine originally present in the sample.[88] This procedure precludes use of the commercially available N-C^2H_3 cocaine as an internal standard for measurement of cocaine because most of the prominent ions in the mass spectrum of N-C^2H_3 cocaine coincide with ions in the spectrum of the O-C^2H_3 cocaine formed by the derivatization. The same research group compared both electron ionization and positive ion chemical ionization of benzoylecgonine derivatized with iodomethane-d_3 vs. derivatization by treatment with hexafluoroisopropanol in the presence of pentafluoropropionic anhydride.[88] The latter procedure converted benzoylecgonine to its hexafluoropropionate ester, which gave EI and CI mass spectra with abundant high-mass ions (EI-MS: m/z 318, 334, and 439; CI-MS: m/z 318 and 440). The detection limits (based on a signal-to-noise ratio of >5) for analysis by positive ion chemical ionization were reported to be 40 ng/ml using the iodomethane-d_3 derivatization and 20 ng/ml when the hexafluoroisopropyl derivative was formed.

Two separate derivatizations were used for the analysis of cocaine, benzoylecgonine, and ecgonine methyl ester in blood and in urine.[98] After solid-phase extraction of the drug and its two major metabolites, the eluent was divided into two equal aliquots. One aliquot was heated with dimethylformamide dipropyl acetal in dimethylformamide to form the propyl ester of benzoylecgonine, while the other aliquot was treated with 4-fluorobenzoyl chloride in pyridine at 85°C for 1 h to convert ecgonine methyl ester to 4-fluorobenzoylecgonine. Cocaine in the extracts was unaffected by either derivatization. GC/MS analysis of the two extracts was performed by monitoring the ion currents corresponding to the molecular ions for each of the derivatives.

Reference standards containing cocaine and benzoylecgonine in freeze-dried urine were analyzed by GC/MS after reconstitution of the urine, solid-phase extraction, and derivatization of the extracted benzoylecgonine by treatment with bis(trimethylsilyl)acetamide (BSA) at 60°C for 1 h.[90] However, trimethylsilylation has more often been achieved by treatment with BSTFA.[96,99]

The t-butyldimethylsilyl ester of benzoylecgonine is less subject to hydrolysis than the trimethylsilyl ester, and can be prepared by derivatization of extracted benzoylecgonine with N-methyl-N-t-butyldimethylsilyl trifluoroacetamide with 1% t-butyldimethylsilyl chloride.[87]

4. GC/MS Conditions and Performance

Most GC/MS assays for cocaine and metabolites have employed either dimethylsilicone or 5% phenylmethylsilicone-coated capillary columns. Typically oven temperatures are programmed from 150 to 280°C at 15°/min; injector and transfer-line temperatures maintained between 250 and 300°C.

The electron ionization mass spectrum of cocaine is shown in Figure 2. Many of the cocaine metabolites show similar fragmentation pathways and therefore are readily identified from their mass spectra even when reference mass spectra are not available.[74,75]

Full-scan spectra of cocaine and cocaethylene extracted from blood and tissue specimens were obtained on an ion trap mass spectrometer at concentrations as low as 5 ng/ml with signal-to-noise ratios exceeding 300 for the molecular ions.[78]

For chemical ionization mass spectrometry of cocaine and its metabolites, methane has served as the reactant gas in several procedures, but the intensity of the MH+ ion is greater when ammonia or isobutane is the reactant gas.[100] Concentrations of 1 ng/ml of cocaine and benzoylecgonine in blood and urine were detected using the isobutane CI procedure, as was 10 ng/ml of ecgonine methyl ester.[92]

With electron ionization, monitoring the ion current at m/z 303 for cocaine is effective and gives linear response over a large dynamic range. At very low concentrations, however, the variability due to low signal intensity manifests itself. For example, monitoring the m/z 182 ion current allowed determination of concentrations as low as 10 ng/ml, a level that was not possible using the m/z 303 ion.[94]

C. Amphetamines

The amphetamines, as a class of drugs, include a variety of closely related compounds. All are simple molecules which, in turn, are closely related to a variety of substances that occur naturally in the body or are found in over-the-counter medications. The potential presence of related compounds in body fluids constitutes the major difficulty in conclusive identification of amphetamines. Derivatization can enhance the chromatographic behavior and mass spectral uniqueness of amphetamine drugs, but does not eliminate interference problems. Furthermore, legitimate use of Vicks Inhaler, which contains l-methamphetamine, makes analysis of enantiomer composition particularly important in the interpretation of forensic results. Although potential pitfalls exist, procedures have been developed that minimize the chances of incorrect identification of the amphetamines.

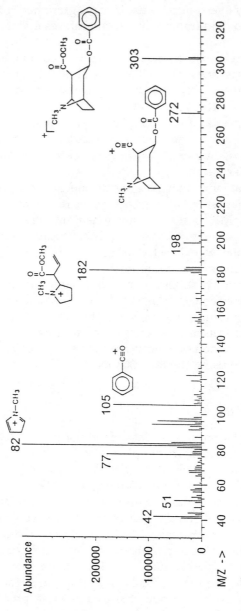

Figure 2.
The electron ionization mass spectrum of cocaine.

Methamphetamine is the most commonly abused of this class of drugs in the U.S. and in Japan. In some European countries there is a greater problem with abuse of amphetamine and/or the synthetic analogs: methylenedioxyamphetamine (MDA), methylenedioxymethamphetamine (MDMA), and methylenedioxyethylamphetamine (MDEA).

The metabolism and excretion of amphetamine and methamphetamine are fairly rapid. Methamphetamine is primarily metabolized to amphetamine, which in turn is metabolized to a variety of minor metabolites, some of which are bioactive, including 4-hydroxyamphetamine, 4-hydroxynorephedrine, and norephedrine. Phenylacetone is also a metabolite and it is further oxidized to benzoic acid, which is then conjugated with glycine to form hippuric acid. Glucuronide conjugates of the benzoic acid and the hydroxylated amphetamines are also excreted in urine.

The proportion of ingested amphetamine and methamphetamine excreted unchanged and the excretion and distribution of the metabolites are strongly influenced by urinary pH. Acidic urine increases the rate of excretion of the parent drugs. Under alkaline conditions, the kidney reabsorbs more of the drug and thereby increases the half-life of the drugs in the body. Since the elimination profiles of amphetamine and methamphetamine reflect a combination of metabolism and urinary excretion, retention of the drugs due to alkaline urinary pH leads to more extensive metabolism. The normal excretion of unchanged drug is approximately 30 and 43% for amphetamine and methamphetamine, respectively.[101,102] The actual amount of unchanged amphetamine excreted can vary from as little as 1% in highly alkaline urine to as much as 74% in strongly acidic urine. Methamphetamine also shows widely differing excretion profiles which range from 2% unchanged methamphetamine in alkaline urine to 76% in acidic urine. The amount of methamphetamine excreted in the form of amphetamine also varies from 7% in acidic urine to as low as 0.1% in strongly alkaline urine.[102]

Analysis of amphetamines continues to be a challenge that requires substantial care and attention. This fact was dramatically exemplified when several laboratories reported methamphetamine in samples which were actually devoid of the drug.[103] These samples were shown to contain high concentrations of either ephedrine or pseudoephedrine. Although the mechanism that led to the false identifications can still be debated, there is no question that the problem was real and steps had to be taken to prevent its recurrence. Fortunately, several preventive measures have been identified that have minimized the likelihood of these errors recurring. For example, urine specimens can be subjected to a simple periodate oxidation which converts ephedrine and pseudoephedrine to compounds that no longer interfere with the detection of methamphetamine.[104] Another safeguard, based on knowledge of the normal metabolism of meth-

amphetamine, is to test for the presence of amphetamine. Since methamphetamine is metabolized to amphetamine, a urine specimen which actually contains methamphetamine should also contain some amphetamine.
Thus, identification of methamphetamine in a sample containing high
concentrations of ephedrine or pseudoephedrine but no amphetamine
should be highly suspect.

Although reporting that a sample contains a drug which is actually
not present (a false positive) is a more serious error, the inability to
identify amphetamine or methamphetamine in samples when they are
present (false negative) is a more common occurrence. This is because of
the number of amphetamine-similar compounds often present in urine
which can co-extract and interfere with the conclusive identification of
amphetamine or methamphetamine.

1. Internal Standards

More deuterium-labeled analogs are available for amphetamine and
methamphetamine than for any other group of commonly abused drugs.
Examples of deuterium-labeled amphetamine currently available include
the following: amphetamine-d_3 and methamphetamine-d_5 (Sigma Chemical Company, St. Louis, MO); amphetamine-d_5 (1-phenyl-$^2H_5$-2-
aminopropane), amphetamine-d_5 (1-phenyl-2-aminopropane-1,2,3,3,3-$^2H_5$),
amphetamine-d_8, methamphetamine-d_5, methamphetamine-d_8 and meth-
amphetamine-d_{11} (Radian, Austin, TX); and amphetamine-d_6 and meth-
amphetamine-d_6 (Alltech Associates, Deerfield, IL). For the analysis of
MDA and MDMA, pentadeuterated isotopomers of these compounds are
also available (Radian). Some procedures use one or the other of these
internal standards for the quantitative analysis of both drugs; however,
the most accurate method is to use the corresponding deuterium-labeled
compound for determination of each analyte.[105] Nondeuterated internal
standards used for analysis of amphetamines include N-propylamphetamine,[106] 4-phenylbutylamine,[106,107] and phenylcyclohexylamine.[107] An advantage of nondeuterated internal standards is that, because they do not
co-elute with the target analyte, they allow full-scan mass spectral analysis
of the samples without the need to invoke complex spectral subtraction
algorithms to obtain pure spectra. In addition, they allow for rapid screening by techniques such as gas chromatography with nitrogen/phosphorus
detection. Any samples in which the screening indicates the presence of
the drug can be confirmed immediately by injection of the same extract
into a GC/MS.[106]

2. Extraction

A wide variety of procedures exist for the extraction of the amphetamines. Perhaps the fastest method for the extraction of amphetamine and

methamphetamine from urine was described by Fitzgerald et al.[108] They added 100 µl of 12 M sodium hydroxide and 100 µl of chloroform to a 2-ml urine sample, mixed the sample, separated the phases, and injected the chloroform layer into the GC/MS. Most extraction procedures are more extensive, requiring more time and effort but giving cleaner extracts. A slightly longer procedure involved extraction with chloroform:isopropanol (9:1) after addition of NaOH to a urine sample. After centrifugation the organic layer was separated and concentrated by evaporation. The extraction efficiencies for this procedure were 53 and 65% for amphetamine and methamphetamine, respectively.[107] Although these percentages are low, the concentrations of amphetamines typically encountered in urine are generally high, and therefore generally do not require higher extraction efficiencies.

Liquid-liquid extraction procedures often employ a back-extraction to achieve further purification. As an example of such a procedure, initial extraction of amphetamines from a 2-ml urine sample (adjusted to a pH ≥ 10) into 1-chlorobutane was followed by back-extraction into an acid solution (0.15 M H_2SO_4). The sample was then made alkaline again by adjustment to pH ≥ 10 with 1.0 ml 1 M NaOH, and extracted into 10 ml of 1-chlorobutane before derivatization and GC/MS analysis.

Solid-phase extraction of amphetamine and methamphetamine has been accomplished using a variety of different sorbent materials. Recoveries have ranged from 50% to essentially 100%. Recoveries of 50 and 70% were reported using ChemElut (Varian) extraction cartridges. The lowest recovery of those reported, this method was designed to extract up to 300 different compounds, including virtually all potentially abused drugs and their metabolites from a single sample, thereby compromising the recovery of single analytes.[8] Recoveries of 78 and 87% for amphetamine and methamphetamine were reported using Detectabuse (Biochemical Diagnostics) extraction cartridges.[106] This study used N-propylamphetamine as the internal standard which was also reproducibly recovered from the columns giving between-run relative standard deviations of approximately 9 and 6% for amphetamine and methamphetamine, respectively. Using the Bond Elut Certify extraction cartridges (Varian), Gan et al. obtained recoveries of 66 and 81% for amphetamine and methamphetamine, respectively.[109] These authors determined that the low extraction efficiencies were due to evaporative loss of amphetamine and methamphetamine rather than poor recovery of the drugs from the column. After elution from the column, the organic extract was concentrated and reconstituted in 1-chlorobutane and washed with 1 M NaOH. This additional step gave significantly cleaner extracts. The method was also effective in the extraction of MDMA from urine with a recovery of 95%.

The highest reported recoveries of amphetamine and methamphetamine using solid-phase extraction were 98 and 100%, respectively, using CleanScreen DAU cartridges (United Technologies, Inc., Bristol, PA).[110]

One way to provide a cleaner extract (regardless of the extraction procedure) and to eliminate the possibility of artifactual "methamphetamine" caused by large amounts of ephedrine or pseudoephedrine is to destroy these compounds prior to GC/MS analysis. This can be accomplished by addition of a small amount of periodate to the urine sample, followed by mixing and a 10-min incubation.[104] This method not only eliminates a potential misidentification problem, but also provides a cleaner extract for injection into the GC/MS.

3. Derivatization

Amphetamine and methamphetamine can be analyzed by GC/MS without derivatization. However, the major ions in their EI mass spectra (m/z 44 and 58) are present in the mass spectra of many other compounds, and therefore are not very useful for structural identification. The lack of unique spectra for amphetamines has led to investigation of various derivatives. Many provide substantial improvement in the mass spectral characteristics of both amphetamine and methamphetamine. For example, Figure 3 compares the EI mass spectra of underivatized amphetamine, methamphetamine, and their respective heptafluorobutyryl (HFB) derivatives. An evaluation of derivatization with heptafluorobutyric anhydride (HFBA) revealed that a 20-min incubation at room temperature using 25 μl of HFBA gave optimal results for up to 50 μg of amphetamine. Increasing the incubation time to 1 h, the temperature to 60°C, or the volume of derivatizing reagent to 100 μl did not increase the yield of derivatized product.[111]

An example of the elution order of the HFB derivatives of some of the amphetamine-type drugs encountered in forensic specimens is shown in Figure 4. The 12 m × 0.20 mm i.d. H-P Ultra 2 capillary column was temperature programmed from 115 to 250°C at 15° per min.

Perfluorinated anhydrides have been used most often for derivatization of amphetamines for GC/MS analysis. However, some investigators have chosen to use less volatile derivatives in order to achieve better separation from co-extractants. For example, the N-trichloroacetyl derivatives are substantially less volatile than the N-trifluoroacetyl derivatives.[112,113] The major limitation of the trichloroacetyl derivatives is the absence of three or more ions well suited to identification by selected ion monitoring.[113] Three separate ions can be monitored for both drugs, but only if the internal standard is carefully selected and at least one of the ions corresponds to a ^{37}Cl isotope. However, monitoring the ^{35}Cl and ^{37}Cl isotope ions of the same fragment ion is not as structurally diagnostic as monitoring ions from two structurally different ions.

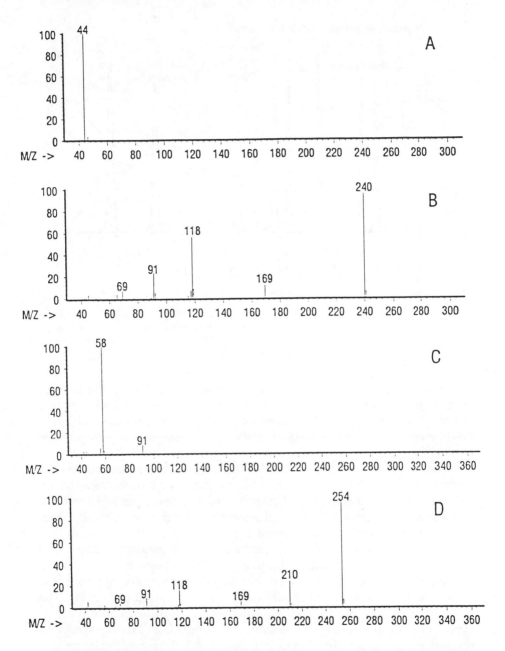

Figure 3.
Mass spectra of (A) amphetamine, (B) HFB-amphetamine, (C) methamphetamine, and (D) HFB-methamphetamine.

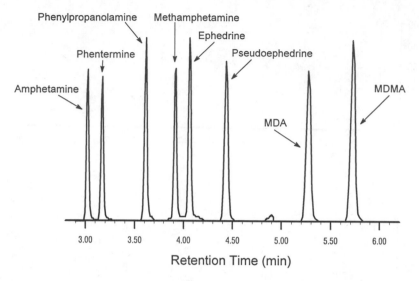

Figure 4.
Ion current profile from GC/MS analysis of the hepatofluorobutyramide derivatives of amphetamine-type drugs. The standard mixture was chromatographed on
a 12-m × 0.20-mm i.d. H-P Ultra 2 capillary column (0.33-μ film) temperature
programmed from 115 to 250°C at 15°/min.

Hornbeck and Czarny have compared the retention times and EI mass
spectra of 11 different derivatives of amphetamine and methamphetamine.[112] More recently, derivatization with perfluorooctanoyl chloride
has been advocated for GC/MS analysis of amphetamines in post-mortem
blood.[105]

Amphetamines can be converted to stable silylated derivatives by
treatment with MTBSTFA with 1% t-butyldimethylchlorosilane. The resulting N-t-butyldimethylsilyl derivatives of amphetamine and methamphetamine are reported to be well separated chromatographically from
other common amphetamine-type drugs.[114]

The EI mass spectra of many derivatives of amphetamine and methamphetamine show an abundant ion at m/z 91. However, this is also an
abundant ion in the EI mass spectra of many endogenous compounds and
therefore it is not well suited for selected ion monitoring analysis. The
need to monitor m/z 91 is eliminated by derivatization with 4-
carbethoxyhexafluorobutyryl chloride.[115] The resulting derivatives contain
abundant high-mass ions at m/z 294, 266, and 248 for amphetamine and
m/z 308, 280, and 262 for methamphetamine. Some of these ions, however,
correspond to fragments of the carbethoxyhexafluorobutyryl substituent,
and are also present in the EI mass spectra of other derivatized primary and
secondary amines. Therefore, they are not as structurally diagnostic as ions
which include the amphetamine portion of the molecule.

Amphetamines, like most drugs, are asymmetric molecules that exist in different enantiomeric forms. This fact typically has more pharmacological than analytical importance, but in the forensic analysis of drugs, identification of the enantiomers can be critical. The *l*-enantiomers of amphetamine and methamphetamine have significantly different pharmacological activity than the *d*-enantiomers and are less subject to abuse. Although amphetamine and methamphetamine are schedule II controlled substances, the Vicks Inhaler sold in the U.S. contains the *l*-enantiomer of methamphetamine and is exempted from control. After use of Vicks Inhaler, *l*-methamphetamine will be excreted in the urine. It is therefore useful to be able to distinguish between the *d*- and *l*-enantiomers of methamphetamine in order to help determine whether methamphetamine in a donor's urine is due to methamphetamine abuse or legitimate application of Vicks Inhaler. One can readily identify the enantiomers of amphetamine and methamphetamine either by using GC columns coated with a chiral stationary phase or by derivatizing the drugs with a chiral reagent. This is most often accomplished by derivatization with trifluoroacetyl-*l*-prolyl chloride (TPC).[108,116,117] On standard GC columns, the enantiomeric forms of amphetamine and methamphetamine are readily separated as the TPC derivative. A typical procedure involves treating the extract with 50 µl of TPC reagent at room temperature for 15 min, washing with 3 ml of 0.01 *M* NaOH (to eliminate excess derivatizing reagent), separating the organic phase, and evaporating it to dryness. The extract can then be reconstituted in ethyl acetate and injected.[117] Formation of the TPC derivative has also been accomplished by an on-column derivatization.[108] In this procedure, 3 µl of the chloroform extract is drawn into a 10-µl syringe, followed by 3 µl of 0.1 *M* TPC; then the contents are injected into the GC/MS. This method has the advantage of allowing rapid determination of enantiomeric composition from the same extract used for quantitative analysis, but has the disadvantage of exposing the system to excess derivatizing reagent, which shortens the life of the GC column. Another approach is to use a derivatizing reagent which can serve both needs with changes in GC conditions. A method using (–)-menthyl chloroformate gave excellent quantitative results but was not able to separate all enantiomers even with long GC run times.[118] An extensive review of methamphetamine enantiomer analysis has been published.[119]

4. GC/MS Conditions and Performance

GC conditions for the analysis of amphetamines can vary considerably. Conditions which are well suited for one derivative may cause significant problems with another derivative. The possibilities for selection of derivatives, GC conditions, and column liquid phases are reflected in the many published procedures, each of which offers one advantage or another over others. Even under the same GC conditions, a change from one derivative to another will change the retention time not only of the drug of interest but also of potentially interfering substances.[120] It has been the experience of one of the authors (JTC) that not one set of assay parameters is sufficient to eliminate all potential interferences in urine samples.[117] However, while no procedure may always be successful, a number of GC conditions serve the majority of samples well. These include both isothermal and temperature-programmed techniques.

Isothermal conditions and retention times for underivatized and 12 different derivatives of amphetamine and methamphetamine have been published along with major ion fragment patterns.[112,115] Procedures using temperature programming ranged over initial oven temperatures of 100 to 180°C, increasing 15 to 30°C/min to final temperatures of 140 to 270°C. For most assays, the upper temperature was set higher than required to elute amphetamine and methamphetamine in order to prevent build-up of less volatile co-extracted compounds in the column. Some injection temperatures were as high as 270°C, while others were held below 200°C. Higher temperatures tend to keep the injection port cleaner, but high temperatures have been associated with the formation of the "methamphetamine" artifact described earlier.

Capillary columns of methylsilicone or 5% phenylmethylsilicone are most commonly used for the analysis of amphetamines, although 50% phenylmethylsilicone has been used to separate co-eluting compounds which could not be resolved by either of the other columns.[117]

Most GC/MS assays for amphetamines have employed electron ionization. However, both positive and negative ion chemical ionization techniques have been applied to analysis of amphetamines.[121–123]

Enantiomer analysis requires different GC conditions from those that are satisfactory for achiral derivatives. Since the use of a chiral derivative may result in two separate peaks from each asymmetric compound, the chromatogram can be more complex and the potential for interference more likely. Conditions which lead to baseline resolution of amphetamine and methamphetamine enantiomers and avoid the interference of virtually all other related compounds generally require longer retention times relative to achiral procedures.

Chiral derivatization is typically invoked to identify the enantiomeric composition of the drugs rather than for qualitative and quantitative analysis of the compounds. In one such method, a 12-m DB-1 column was maintained at 120°C for 2 min; then the temperature was increased to 200°C at 4°C/min. This procedure gave retention times for the enantiomers of approximately 14 and 18 min for amphetamine and methamphetamine, respectively. Although this was a relatively long run time, the enantiomer peaks were well resolved even at high drug concentrations and interference from other related compounds was minimized.[117] Figure 5 shows an example of such a chromatogram. Shorter retention times can be achieved by changing the GC conditions. With an initial temperature of 190°C for 4 min, then increasing to 250°C at 20°C/min, retention times for the methamphetamine enantiomers were less than 5 min; however, the methamphetamine enantiomers were not baseline resolved and no retention times were given for the amphetamine enantiomers.[108] Assessing the amphetamine enantiomers can be important for some samples.[117]

D. Opiates

The opium poppy plant contains alkaloids that have been used and abused for centuries. The most abundant of these alkaloids are morphine and codeine. Heroin, the most widely abused opiate today, is prepared by acetylation of morphine. Heroin is rapidly metabolized to 6-acetylmorphine and subsequently to morphine. Because virtually no intact heroin is excreted in the urine, identification of heroin abuse by analysis of urine samples is often difficult. A heroin user's urine normally contains high concentrations of morphine, but morphine in urine can also result from legitimate ingestion of codeine-containing cough medicines or pain relievers, or from ingestion of bakery goods containing poppy seeds.[124–128] Detection of 6-acetylmorphine in a urine sample is considered conclusive evidence for heroin use, but this metabolite can only be detected in the urine for a brief time after heroin administration.[129]

Heroin is not only short-lived in the body, but it is unstable *in vitro*. Therefore, concentrations of heroin and 6-acetylmorphine can be affected by the way in which the samples are handled following collection. The degradation of heroin to 6-acetylmorphine occurs spontaneously as well as enzymatically. Enzymatic hydrolysis can be controlled by addition of enzyme inhibitors, and the spontaneous degradation can be controlled by freezing the sample soon after collection.

Analysis of body fluids for morphine and codeine is further complicated by the number of other opiate drugs which can yield positive results with opiate immunoassays. Well designed GC/MS assays can distinguish between these compounds, either chromatographically and/or by accurate measurement of the relative intensities of structurally diagnostic ions. Some problems do exist however; for example, the trimethylsilyl (TMS) derivatives of morphine, hydromorphone, and norcodeine (a metabolite of codeine) have similar retention times and share the same major ions. A similar problem exists in distinguishing between the TMS derivatives of codeine and hydrocodone. Careful analysis on a well-maintained and controlled instrument can permit reliable identification of these derivatives, but proper selection of the ions to monitor and careful attention to ion ratios are critically important.

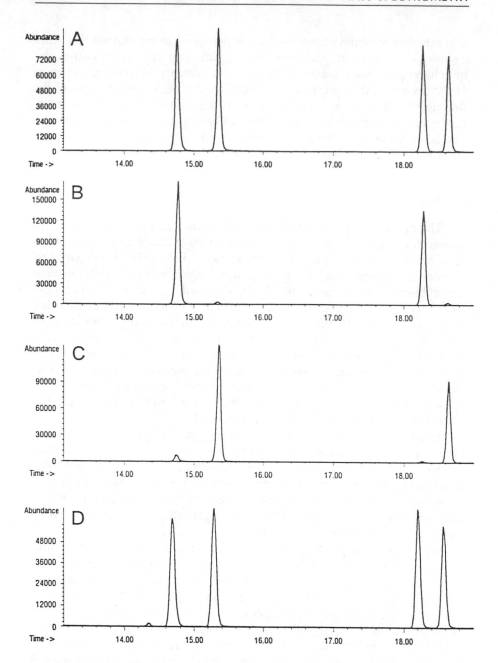

Figure 5.
Ion current profiles of the N-trifluoroacetyl-l-prolyl derivatives of d,l-amphetamine and d,l-methamphetamine (A), l-amphetamine and l-methamphetamine (B), d-amphetamine and d-methamphetamine (C), and d,l-amphetamine-d$_5$ and d,l-methamphetamine-d$_5$.

The GC/MS analysis of opiates has been recently reviewed.[68]

1. Internal Standards

Commercially available deuterium-labeled opiates include: morphine-d_3, codeine-d_3, (Alltech Associates, Deerfield, IL; Radian, Austin, TX; Sigma Chemical Company, St. Louis, MO), oxycodone-d_3 (OC^2H_3), hydrocodone-d_3, hydromorphone-d_3, and 6-acetylmorphine-d_3 (Radian, Austin, TX).

Although not commercially available, heroin-d_9 [3,6-di-(O-$^2H_3$-acetyl)-15,15,16-$^2H_3$-morphine] was synthesized for a study that measured heroin, acetylmorphine, and morphine concentrations in urine and saliva.[130] Use of the deuterated isotopomer as the internal standard compensated for the hydrolysis of heroin in samples which, even under the best of conditions, was approximately 5%.

Each of the deuterium-labeled compounds listed above offers substantial advantages over nondeuterated compounds as an internal standard for the corresponding drug. The solubilities and pK_as of morphine and codeine are sufficiently different that using a single internal standard for quantitation of both drugs is not recommended. For example, at some pH values the extraction efficiencies of these compounds vary significantly. At low pH (<5), morphine recovery is proportionately higher than codeine recovery, while the reverse is true at pH levels greater than 8.[131] The same potential problem exists when other opiates, such as nalorphine, are used as internal standards for measurement of morphine and/or codeine.[132]

2. Hydrolysis and Extraction

Liquid-liquid and solid-phase extractions have both been used successfully to analyze morphine and codeine in biological samples. As the solubilities of morphine and codeine differ, extraction methods must accommodate this difference. Often the extraction efficiency of one of the two drugs suffers if a highly selective solvent system is used; if the solvent system chosen extracts both drugs with high efficiency, unacceptable amounts of potentially interfering substances may be co-extracted.

Most of the morphine and codeine excreted in urine is conjugated and requires hydrolysis prior to derivatization and GC/MS analysis.[133] Conjugation occurs primarily with glucuronic acid, but also to some extent with sulfate;[134] therefore, complete enzymatic hydrolysis requires a combination of glucuronidase and sulfatase. Enzymatic hydrolysis typically includes addition of the enzyme to the sample, followed by incubations ranging from several hours to overnight.[135] Acid hydrolysis includes acidification of the sample followed by heating at ≥100°C for 15 to 30 min. Acid hydrolysis dissociates all conjugates, but unfortunately it will also convert any heroin and acetylmorphine present in the sample to morphine.

In a study comparing acid hydrolysis of opiates with enzymatic hydrolysis, both procedures were found equally effective.[136] Acid hydrolysis was compared to the addition of 10,000 units of β-glucuronidase from *Patella vulgata* to each 2-ml sample after adjustment to pH 6.0 to 7.5, followed by incubation at 60 to 65°C for 2 h. Analysis of 20 batches over 14 weeks showed coefficients of variation 5.5 and 2.7%, respectively, for enzymatic vs. acid hydrolysis.

After an extensive evaluation of experimental parameters for hydrolysis of morphine glucuronide by β-glucuronidase from *E. coli,* the investigators settled on the following conditions. The crystalline β-glucuronidase was weighed and reconstituted in 0.075 M phosphate buffer (pH 6.8). To each 1-ml urine sample was added approximately 1200 Fishman units and the sample was incubated at 40°C for 1.5 h.[137]

In a reported liquid-liquid extraction procedure performed after acid hydrolysis, the pH of the urine was adjusted to 9.0 to 9.3 with sodium hydroxide and ammonium chloride. After centrifugation, the extract was treated with methylene chloride:isobutanol (9:1), washed with phosphate buffer, back-extracted into acid, alkalinized with sodium carbonate, and re-extracted with methylene chloride:isobutanol. Recoveries of the drugs in this extraction were 40 and 58% for morphine and codeine, respectively. Although these recoveries were not high, the extracts were reasonably free from interferences.[138]

An extensive evaluation of solid-phase extraction of morphine and codeine on C-18 sorbent material revealed several important considerations. A pH outside the range of 5 to 8 gave substantially lower recoveries. Solvent washes also had a significant impact, particularly with respect to the use of alcohols.[131] The study demonstrated that a final wash with methanol was superior to washing with less polar alcohols such as ethanol and isopropanol.

Extraction of morphine from vitreous humor with C-18 reverse-phase adsorbents gave cleaner extracts than either solid-phase extraction with normal-phase packing material or liquid-liquid extraction.[139]

Pretreatment of blood samples before their application to a solid-phase extraction column was shown to have a significant impact on drug recovery; direct application of blood caused the columns to become plugged. Precipitation of the blood with methanol or acetonitrile and application of the supernatant to the extraction column gave poor extraction efficiencies. However, sonication of diluted blood before application to the column at low pH (3.3) and elution with methanol:ammonia (98:2) consistently yielded greater than 80% recovery.[140]

3.   Derivatization

A wide variety of acylating and silylating reagents have been used for derivatization of opiates for GC/MS analysis. In studies comparing the stabilities of various derivatives of morphine and codeine,[132,138,141] the acetyl derivative was found to be more stable than the others evaluated. However, since acetic anhydride converts both morphine and 6-acetylmorphine to diacetylmorphine (heroin), it cannot be used for assays intended to distinguish between these compounds, unless deuterium-labeled acetic anhydride is used.[142] Propionic anhydride also converts opiates to relatively stable derivatives and permits distinguishing between morphine, monoacetylmorphine, and heroin.[137]

Figure 6 illustrates the elution order of commonly encountered opiate-type drugs after derivatization with deuterated acetic anhydride. The standard mixture was chromatographed on a 12 m × 0.25 mm i.d. DB5 capillary column which was temperature programmed from 185 to 320°C at 20°/min.

Derivatization of morphine and codeine with perfluoronated anhydrides, including trifluoroacetic anhydride (TFA), pentafluoropropionic anhydride (PFPA), and heptafluorobutyric anhydride (HFBA), has been evaluated in several studies. Some disagreement exists concerning the utility of these derivatives and the results have varied considerably. In a study by Paul et al., TFA, PFPA, and HFBA gave acceptable chromatography, but instability was a problem.[138] Partial spontaneous hydrolysis of derivatives occurred within 24 h. In other studies, the chromatography, particularly of codeine, was poor;[132,141] yet Fuller and Anderson found no such problems for the TFA derivatives of free morphine, codeine, or acetylmorphine.[143] Procedures using nalorphine as the internal standard exhibited problems with quantitation due to differing stabilities between the drug and the internal standard;[132] however, use of deuterium-labeled isotopomers essentially eliminated this problem.[143] Trimethylsilyl derivatives of the opiates are also unstable due to their susceptibility to hydrolysis in the presence of even small amounts of moisture. This instability can be overcome by leaving the derivatized extract in an excess of derivatizing reagent rather than evaporating and reconstituting the extracts before injection into the GC/MS.

Derivatization of extracts which contain hydromorphone and hydrocodone can present problems with identification. A number of derivatizing reagents react with the enol form of these drugs to give spectra very similar to those of derivatized morphine and codeine. Mild derivatization conditions and alternative derivatizing reagents, such as acetic anhydride or N-trimethylsilylimidazole, minimize the formation of enol derivatives. Although most common for TMS derivatives, the interference caused by hydromorphone and hydrocodone enol forms has also been reported in a study where derivatives made with pyridine and acetic anhydride were heated to 70°C for 20 min,[141] but was not observed under milder conditions.[132,138]

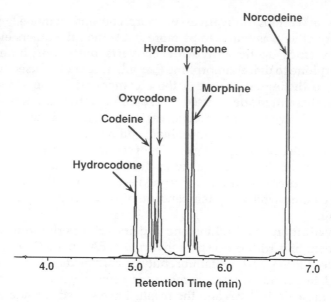

Figure 6.
Ion current profile from GC/MS analysis of the derivatives of opiate drugs after derivatization with deuterated acetic anhydride. The standard mixture was chromatographed on a 12-m × 0.25-mm i.d. DB5 capillary column temperature programmed from 185 to 320°C at 20°/min.

4. GC/MS Conditions and Performance

Both isothermal and temperature-programmed conditions have been used for GC/MS analyses of morphine and codeine. For example, a constant oven temperature of 240°C was used for the analysis of the acetyl derivatives of codeine and morphine.[138] Temperature programs vary considerably depending on the derivative used, but typically involve initial temperatures of 150 to 180°C, increasing at 10 to 40°C/min to final temperatures of 270 to 300°C. In an extensive study of morphine, codeine, and a number of different metabolites, several different stationary phases were compared. Although none of the columns was able to completely separate all of the drugs and their metabolites, the more polar columns gave better overall performance than the less polar ones.[144] For analysis of morphine and codeine in the absence of multiple metabolites, most assays utilize methylsilicone or 5% phenylmethylsilicone as the stationary phase.

Analysis of acetylmorphine has been accomplished using procedures specifically designed for the purpose as well as procedures that are general for opiates. Detection limits for 6-acetylmorphine in urine as low as 0.8 ng/ml have been reported.[145] The ability to detect acetylmorphine at levels below 10 ng/ml, a relatively high concentration for this metabolite, is important. Using pentafluoropropionate derivatives of morphine, codeine, and acetylmorphine, Schuberth and Schuberth were able to detect 0.5 ng of acetylmorphine per gram of post-mortem blood.[146]

Most published GC/MS assays for opiates involve electron ionization. GC/MS/MS analyses and chemical ionization are less often used, but are routinely performed in a few toxicology laboratories. Chemical ionization of ten opiate alkaloids was evaluated using methane, isobutane, and ammonia as reactant gases.[147]

The specificity of GC/MS/MS analysis of the pentafluoropropionate derivative of morphine permitted use of a simple one-step extraction of the drug from blood. The method involved selected reaction monitoring of two parent-to-product ion transitions. The method gave a 5:1 signal-to-noise ratio for the ion current resulting from the $577 \rightarrow 414$ transition in the analysis of a sample containing 1 ng/ml (2-ml sample) of morphine, while the same sample produced no discernible peak using GC/MS with selected ion monitoring.[148]

E. Barbiturates

The legitimate medical use and the abuse of barbiturates have both been declining in recent years. Also, in spite of the large number of different barbiturates that are marketed, only a few are commonly identified in forensic samples. Perhaps for these reasons, relatively few new GC/MS assays for barbiturates have been published during the past 8 years; assays reported earlier were previously reviewed.[1,2]

Barbiturate concentrations in urine following recent ingestion are generally quite high, so achieving adequate sensitivity is seldom a problem. The major difficulties encountered in the GC/MS determination of barbiturates in body fluids are (1) identification of the specific barbiturate(s) present in a sample and (2) achieving good chromatographic performance. As some barbiturates give nearly identical mass spectra, identification often requires comparison of the drug's retention time with that of a reference standard. Underivatized barbiturates, particularly phenobarbital, have a strong tendency to give broad and unsymmetrical chromatographic peaks unless the GC column is well deactivated. Derivatized barbiturates generally show much improved chromatographic behavior.

1. Internal Standards

Several deuterium-labeled barbiturates are commercially available (Radian Corporation, Austin, TX). However, if they are to be used as internal standards in GC/MS assays employing electron ionization, it is important to confirm that the deuterated barbiturate gives an EI mass spectrum containing major ion peaks that retain at least two of the deuterium labels. For example, two different deuterium-labeled phenobarbital isotopomers are available from Radian Corporation, one containing five deuterium atoms attached to the ethyl group and one containing five deuterium atoms attached to the phenyl group. The base peak in the EI mass spectrum of the former is at m/z 205 which is only 1 mass unit different from the base peak (m/z 204) in the EI mass spectrum of unlabeled phenobarbital. In contrast, the base peak in the EI mass spectrum of the phenobarbital containing deuteriums attached to the phenyl ring is at m/z 209. Clearly the phenyl-labeled isotopomer is the preferred internal standard for EI assays. With methane or ammonia chemical ionization (CI), both isotopomers give abundant protonated molecule peaks at m/z 238; therefore, either labeled analog is a suitable internal standard for CI assays for barbiturates.

Many of the GC/MS assays described in the literature employ as internal standards barbiturates that are not used medically, such as hexabarbital[149] and tolylbarbiturate.[150]

2. Extraction

Barbiturates can be extracted from urine with a mixture of hexane and ethyl acetate (6:4 v/v) after the pH is adjusted to 5.5 with a phosphate buffer.[150] The extraction efficiencies range from 75 to 84%. The following unpublished procedure, used in the laboratory of one of the authors (RLF), includes a back-extraction cleanup step. After acidification of the urine with 0.5 M hydrochloric acid, the urine is extracted into methylene chloride. The barbiturates are back-extracted into 0.1 M sodium hydroxide. The aqueous alkaline extract is then acidified and the barbiturates are re-extracted into methylene chloride. GC/MS analysis of the resulting extracts, performed without derivatization, gives relatively clean ion current profiles for each of the monitored ions. It is important, however, to minimize the length of time that the barbiturates are exposed to the dilute alkali, since basic pH can cause degradation of the drugs.

A liquid-liquid extraction was reported in a GC/MS analysis for phenobarbital, phenobarbital labeled with two ^{15}N and one ^{13}C, and their corresponding p-hydroxylated metabolites in plasma and urine.[151] After addition of deuterated phenobarbital as the internal standard and an acetate buffer (pH 5.2), the plasma was extracted with chloroform-isopropanol (20:1). Urine samples were treated with H. pomatia β-glucuronidase before extraction, to hydrolyze the conjugated metabolite.

Radiolabeled phenobarbital was used in a comparison of different procedures for extraction of drugs from urine.[152] Ten-ml aliquots of urine containing 10 µg/ml of the radiolabeled phenobarbital were extracted by the following procedures: liquid-liquid extraction with methylene chloride at pH 5.2 or at pH 9.3, solid-phase extractions using either Sep-Pak or

Baker C-18 cartridges, or columns containing either Amberlite XAD-2 resin or diatomaceous earth Celite 560. Because of its acidity, phenobarbital was poorly extracted at pH 9.3. At pH 5.2, 3 sequential extractions, each with 5 ml of methylene chloride, gave a total recovery of 98%. The XAD-2 columns were eluted with 20-ml portions of methanol, the Celite 560 columns with 10-ml portions of methylene chloride, the Baker C-18 columns with 1-ml portions of acetone:methylene chloride (1:1), and the Sep-Pak columns with a 2-ml portion of 20% methanol in water followed by elution with 2-ml portions of methanol. The amounts of radiolabeled phenobarbital in each of the initial elution volumes were XAD-2, 89%; Celite 560; 34%; Sep-Pak, 86%; and Baker C-18, 80%. Extraction with the C-18 bonded silica cartridges was judged to be the least time-consuming, due to the small volumes of solvent required for the extractions and the ease with which the solvents could be removed by evaporation from these extracts.

A solid-phase extraction procedure specifically designed for confirmation of barbiturates in urine was recently reported.[149] A sodium acetate buffer (pH 7.0) was added to a 5-ml urine sample and the pH adjusted to between 5 and 7. A Bond Elut Certify II column was conditioned by washing with 2 ml of methanol and 2 ml of sodium acetate buffer. The urine sample was then applied to the column, the column was washed with buffer and with hexane:ethyl acetate (20:1), and the barbiturates were eluted with 2 ml of hexane:ethyl acetate (3:1). The percent recoveries for five barbiturates ranged from 82 to 100%.

3. Derivatization

Derivatization by alkylation can improve the gas chromatographic and mass spectrometric characteristics of barbiturates. Published accounts describe various tetraalkylammonium hydroxide reagents for on-column derivatization and GC analysis of 16 barbiturates.[153] A similar on-column derivatization employing a proprietary methylation reagent (Barb-Prep, Alltech Associates, Deerfield, IL) was used for confirmation and quantitation of barbiturates in urine by GC/MS.[150]

In another report, a different alkylating reagent was used for GC/MS determination of barbiturates in blood, urine, and vitreous humor.[154] Ethyl acetate extracts of biological samples were treated with equal amounts of dimethylformamide and dimethylformamide dipropyl acetal, and placed in a 130° heating block for 2 min. After cooling, acetate buffer (pH 7.3) and hexane were added to the derivatization mixture. The hexane layer was removed and analyzed by selected ion monitoring. The investigator in this study preferred the dipropyl acetal reagent to other dimethylformamide acetals because it was the purest of the dimethylformamide acetal reagents commercially available.

4. GC/MS Conditions and Performance

Maurer has published a GC/MS screening procedure for identification of barbiturates and metabolites in urine.[155] The method consists of acid hydrolysis, adjustment of the pH to between 8 and 9, and extraction with methylene chloride:isopropanol:ethyl acetate (1:1:3). The extracts are then acetylated with acetic anhydride in pyridine (3:2) to derivatize polar metabolites. Under these conditions most of the parent barbiturates survive unchanged and can be detected by generating reconstructed ion current profiles for the ions at m/z 83, 117, 141, 169, 207, 221, and 235. The identifications can be confirmed by visual comparison of the full EI mass spectra with reference spectra. The report of this procedure included EI mass spectra for 24 barbiturates and 13 other hypnotics.

A nonchromatographic tandem mass spectrometric method was developed for rapid detection of barbiturates in serum.[156] Citrate buffer (pH 3) was added to serum and the mixture poured into an Extrelut 3 column; the column was extracted with chloroform. The chloroform extract was evaporated to dryness and the residue was introduced into the ion source of a triple-quadrupole mass spectrometer via a direct-insertion probe. Barbiturates were detected by constant "neutral loss scans" corresponding to loss of HNCO; subsequent identification was achieved by "product" and "parent ion" scans. Most barbiturates could be detected at serum concentrations as low as 1 μg/ml. The method is not intended for confirmation or quantitative analysis, either of which would be difficult without chromatographic separation.

Of the recently reported GC/MS assays for barbiturates, the best sensitivity (LOD = 20 ng/ml) was achieved by the method employing liquid-liquid extraction, on-column methylation, chromatographic separation on a 12.5-m × 0.20-mm capillary column (0.33-μm methylsilicone film thickness), and electron ionization with selected ion monitoring.[150]

The fragmentation of barbiturates following either electron ionization or chemical ionization was discussed in an earlier review.[2]

F. Phencyclidine (PCP)

Analysis for phencyclidine (PCP) has proven to be one of the least trouble-prone GC/MS drug assays performed in toxicology laboratories, even though it is often measured at levels lower than those for many other drugs. The drug is easily extracted from biological samples by liquid-liquid or solid-phase methods and requires no derivatization. Abuse of PCP is less common now than in the past, and its use tends to be geographically regionalized. A synthetic drug, PCP is listed under schedule II of the Controlled Substances Act. Abuse can take the form of PCP alone or in combination with another drug (i.e., smoking marijuana sprinkled with PCP). Routes of administration include oral ingestion, intravenous injection, smoking, and insufflation of the powdered form.

PCP metabolism includes hydroxylation of the cyclohexane and piperidine rings to form 4-phenyl-4-piperidinocyclohexanol and 1-(1-phenycyclohexyl)-4-hydroxypiperidine, which are excreted as glucuronide conjugates, and formation of an amino acid metabolite, 5-(N-(1'-phenylcyclohexyl)amino)pentanoic acid. Most of the parent drug and metabolites are excreted in urine; studies have shown that an average of 77% of an intravenous dose is excreted in 10 days. Urine pH has a substantial effect on the rate of excretion of this drug, and acidification of urine has been used in treatment of overdoses.[157]

1. Internal Standards

Deuterated phencyclidine (PCP-d_5) is available from several U.S. sources including Alltech Associates, Deerfield, IL; Radian, Austin, TX; and Sigma Chemical Company, St. Louis, MO. A variety of quantitative procedures have been reported for the analysis of PCP, most of which use the deuterated internal standard.

2. Extraction

Little has changed in methods for analysis of PCP over the past several years other than the development of procedures for solid-phase extraction. Liquid-liquid extractions have been used successfully for many years, and they are relatively simple due to the simplicity of the PCP molecule. Addition of 50 µl of 50% NaOH to a 0.2-ml urine sample, which was then brought to a final volume of 1 ml with distilled water and extracted with 3 ml of chloroform:isopropanol (90:10), proved to be a rapid and effective extraction procedure with a recovery of 87%.[107]

Solid-phase extraction of PCP has been described by a number of groups. An extraction with Clean Screen cartridges (United Technologies, Inc., Bristol, PA) was reported to give essentially 100% recovery of the drug and to co-extract fewer compounds than liquid-liquid procedures.[158] The limit of detection for a GC/MS assay using this extraction was less than 1 ng/ml of PCP in urine. Similar performance was reported using Prep-Sep C-18 cartridges (Fisher Scientific) which yielded recovery of 93% with no interfering substances evident.[159] This procedure was effective not only for extraction of parent PCP but of its acid metabolite as well.

A potential problem is associated with the evaporation step between extraction and injection of the sample into the GC/MS. Loss of PCP can occur if the evaporation temperature is higher than 30°C.[158] Higher temperatures (40°C) can be tolerated, however, if one is careful not to allow the sample to remain at elevated temperatures after evaporation is complete.[160]

3. GC/MS Conditions and Performance

The mass spectrum of PCP shows some interesting characteristics. One is the fact that two of the most abundant ions differ by only 1 m/z. It is generally not considered good practice to monitor ions only 1 m/z apart, partly because all organic ions have a ^{13}C-isotope peak 1 m/z higher. In the case of PCP, two distinct ion fragments give rise to the ions at m/z 242 and 243; the spectrum of the deuterated analog of PCP (phenyl-d_5-PCP) shows a corresponding ion that differs by 2 m/z. This clearly indicates that the loss of deuterium from the molecular ion (m/z 248) of PCP-d_5 gives rise to the ion at m/z 246. It also demonstrates that the loss is from the aromatic ring since that is the location of the deuterium atoms. See Figure 7 for the structure of PCP and its deuterated analog, along with their mass spectra.

Commonly monitored ions for PCP are 243, 242, and 200. These ions are quite stable and provide chromatograms that are relatively free of interferences. Some laboratories have chosen to monitor the less abundant m/z 186 ion rather than m/z 242.[158]

The following is a representative example of the conditions used for analysis of PCP. A 12-m Hewlett-Packard Ultra-2 capillary column was maintained at 120°C for 1 min after injection, the oven temperature was then increased at 40°C/min to 190°C, and held for 6 min. Under these conditions PCP eluted at approximately 5.5 min.[158]

The utility of monitoring one of the metabolites of PCP was evaluated as a way to improve detection of PCP use. Analysis for the metabolite 5-(N-(1'-phenylcyclohexyl)amino)pentanoic acid proved to be straightforward: after extraction, the metabolite was methylated and analyzed by GC/MS. Ions were monitored at m/z 289, 246, and 159.[159] Samples containing PCP often contained higher concentrations of the metabolite than of the parent drug.

Thermal degradation of PCP has been documented at temperatures over 150°C and substantial decomposition can occur at temperatures above 200°C.[161] Currently employed procedures often involve temperatures above 150°C and still provide valid data. One of the major advantages of deuterium-labeled analogs as internal standards is that they behave as their nonlabeled counterparts do. Therefore, if degradation of PCP occurs at these high temperatures, both the drug and its isotopomer will decompose in a proportionate manner, and thus the degradation will not affect quantitative accuracy so long as the signal intensities are adequate.

In a recent study, PCP was shown to bind tightly to hair following environmental exposure.[162] The issue of contamination of hair samples by environmental exposure is a continuing concern when only the parent drug is detected in hair.

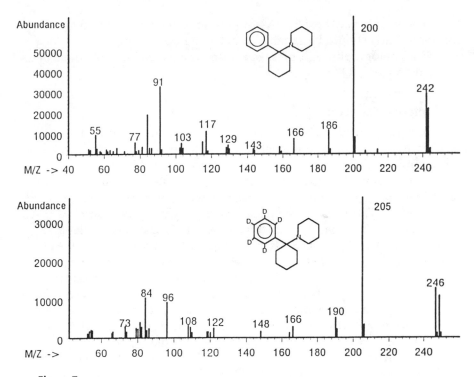

Figure 7.
Structure and mass spectra of PCP and PCP-d$_5$.

G. Lysergic Acid Diethylamide (LSD)

Use of LSD is increasing again in the U.S.[163] and this has generated renewed interest in development and application of methods for detection of LSD use by analysis of body fluids. Conclusive identification of LSD in body fluids is a challenging analytical task. The usual oral dose of LSD is only 20 to 80 µg,[164] and the drug is extensively metabolized, primarily to as yet unidentified metabolites. Furthermore, LSD is sensitive to ultraviolet light and elevated temperatures, and it is relatively unstable in solution at pHs below 4.0.[165] However, the instability of LSD is not a major analytical problem if reasonable precautions are taken in the handling and storage of standards and specimens containing the drug. A far greater problem for analytical laboratories is the strong tendency for LSD and derivatized LSD to undergo adsorptive losses when subjected to gas chromatography; this behavior often prevents detection of the drug at the subnanogram/milliliter concentrations normally encountered in body fluids from LSD users.

Iso-LSD, a nonpsychoactive diastereoisomer of LSD, is often present in illicit preparations. Interconversion of the two isomers can occur during storage, particularly at elevated pH, with the equilibrium favoring conversion of iso-LSD to LSD. Assays for detection of LSD use by analysis of body fluids should be capable of detecting both isomers.

Chromatographic and mass spectrometric assays for LSD have been recently reviewed.[166,167]

1. Internal Standards

A variety of LSD analogs have been used successfully as internal standards. Two deuterium-labeled analogs are commercially available: one contains three deuterium atoms attached to the N-methyl group (Radian, Austin, TX), and other contains six deuteriums attached to the terminal methyls of the diethylamide group (Sigma Chemical Co., St. Louis, MO). The isomeric lysergic acid methylpropylamide (LAMPA), which also makes a very satisfactory internal standard, is available from several vendors in the U.S. (Radian, Sigma Chemical Co., and Alltech Associates, Inc., Deerfield, IL). LAMPA has several advantages as an internal standard. Its extraction and chromatographic and mass spectral characteristics are similar to those of LSD, yet it is easily resolvable from LSD by capillary chromatography; consequently, either full-scan or selected ion monitoring analysis is possible. Also, with LAMPA as the internal standard, the three most abundant high-mass ions (m/z 395, 293, and 253) can be monitored as long as the LAMPA-TMS is chromatographically separated from the LSD-TMS. If N-C^{2}H$_3$-LSD is the internal standard, the m/z 253 ion cannot be used because it does not retain the N-C^{2}H$_3$ group. Paul and co-workers have postulated structures for all of the most abundant ions in the EI mass spectrum of LSD-TMS.[165] In these postulated ion structures, only the molecular ion retains the diethylamide group. Consequently, LSD deuterated on the diethylamide group would not be a good choice as the internal standard for analysis of LSD as the trimethylsilyl derivative.

2. Extraction

LSD can be extracted efficiently from alkalinized urine (pH >9) with organic solvent systems of low polarity, such as 1-chlorobutane,[165,168] or toluene:methylene chloride (7:3 v/v).[169,170] If a very selective method of analysis is available, such as GC/MS with negative-ion chemical ionization[169] or GC/MS/MS,[170] no further sample cleanup is necessary. Methods based on GC/MS with electron ionization generally require additional sample cleanup steps, such as a back-extraction into acid[165] or purification by solid-phase extraction.[165] Published methods for extraction of LSD from blood or plasma also require additional sample cleanup steps.[170,171]

3. Derivatization

To achieve the sensitivity needed for measurement of LSD in biological samples, most published GC/MS procedures have included derivatization of the indole nitrogen to improve the compound's chromatographic behavior. Reduction of LSD's amide group with lithium aluminum hydride is another way of improving its chromatographic behavior,[172] but this approach has not yet been applied to analysis of LSD in biological samples.

Trimethylsilylation has been the method most often used for derivatization of LSD for GC/MS analysis; this can be easily accomplished by heating an extract residue with BSTFA at 70°C for 15 to 30 min. Although the TMS derivative of LSD is fairly stable in the presence of excess silylating agent, attempts to remove the derivatizing reagent before injection of the extract into the GC/MS have been unsuccessful.

Formation of the N-trifluoroacetyl (TFA) derivative of LSD has been used in GC/MS assays employing negative-ion chemical ionization for determination of LSD in urine,[169] and in plasma.[171] After various methods of attaching perfluoroacyl groups to the LSD molecule were investigated, the procedure that proved most effective consisted of heating the extract residue with 20 µl of 30% trifluoroacetylimidazole in toluene and 10 µl of 1% 1,4-dimethylpiperazine in toluene for 20 min at 80°C. The pentafluoropropionate and heptafluorobutyryl derivatives of LSD were similarly prepared, but gave less satisfactory negative-ion chemical ionization mass spectra than the TFA derivative.

4. GC/MS Conditions and Performance

Experience in several laboratories has shown that successful determination of LSD in biological specimens requires a good quality, well-deactivated capillary GC column and a clean, well-deactivated injector insert. In addition, various procedures have been reported for priming the chromatographic column before beginning analysis of LSD samples; for example, successive injections of silylating agent,[168] or simply four or more injections of a derivatized extract of drug-free urine.[165] Other techniques that are reported to minimize adsorptive loss of LSD are (1) reconstitution of the extract residue in 0.001 M triethylamine in ethanol[165] and (2) addition of 1 µg of methysergide to samples before extraction to act as a carrier, and co-injection of the derivatized extract with 0.5 µl of 1% tributylamine.[169]

Typical gas chromatographic conditions for determination of LSD in derivatized extracts consist of splitless injection into a methylsilicone or 5% phenyl methylsilicone capillary column and heating the column from 200 to 300°C at 15°/min. An injection port temperature of 250°C was found to be optimum in one study,[168] but other investigators have used injection-port temperatures as high as 295°C without evidence of thermal degradation of trimethylsilylated LSD.[170]

Most published methods for GC/MS analysis of LSD in biological specimens have employed electron ionization of the N-trimethylsilyl derivative of LSD. All of these methods have used selected ion monitoring of the molecular ion and a few prominent fragment ions in order to achieve the required sensitivity. Even then, interferences by co-extractants can seriously limit an assay's sensitivity unless a highly selective extraction is used. For example, the double extraction procedure employed by Paul and co-workers has permitted measurement of LSD in urine speci-mens at concentrations as low as 100 pg/ml.[165] Other published EI-MS assays employ simpler, less selective extraction procedures, but as some of these assays do not appear to have been applied to the analysis of physi-ological specimens, the sensitivities claimed may not be achievable when applied to analysis of specimens from LSD users.[173,174]

The negative-ion chemical ionization mass spectrum of the N-trifluoroacetyl derivative of LSD contains only one abundant ion, the molecular anion at m/z 419, and its intensity is very dependent on the ion-source temperature.[169] At low ion-source temperatures (<120°C), the trifluoroacetyl derivative of LSD is efficiently ionized, permitting mea-surement of LSD in urine extracts down to 50 pg/ml. The N-demethyl metabolite of LSD ("nor-LSD") forms a bis-(trifluoroacetyl) derivative which is even more efficiently ionized by electron-capture negative ion chemical ionization, and gives an abundant molecular anion at m/z 501. However, a disadvantage of the negative-ion chemical ionization method is that the TFA derivatives of LSD and iso-LSD are difficult to separate chromatographically with the silicone-coated capillary columns in com-mon use.

The most sensitive and specific assay for LSD reported to date is based on gas chromatography and tandem mass spectrometry (GC/MS/MS).[170] The method consists of a simple liquid-liquid extraction, derivatization with BSTFA, and selective reaction monitoring of three product ions formed by collision-induced dissociation of the protonated molecule formed by positive-ion chemical ionization of the trimethylsilyl derivatives of LSD, iso-LSD, and LAMPA. Ammonia is more effective than methane as the reagent gas. When the assay was applied to the analysis of more than 200 urine and plasma specimens that were RIA-positive for LSD, the resulting ion current profiles were essentially devoid of any peaks due to non-LSD related compounds. A limit of detection of 10 pg/ml and a limit of quantitation of 50 pg/ml was achieved. For GC/MS/MS determination of N-demethyl LSD, the best sensitivity has been achieved by collision-induced dissociation of the molecular anion formed by negative-ion chemi-cal ionization of the bis-(trifluoroacetyl) derivative.

H. Benzodiazepines

The benzodiazepines as a class include some of the most frequently prescribed drugs. However, because so many different benzodiazepines are marketed and their potencies vary so widely, development of GC/MS assays for identification and quantitation of these drugs in physiological specimens has challenged many toxicologists. The task has been made easier by the availability of compilations of mass spectra of benzodiazepines and their metabolites,[175–183] and by recent reviews of methods for chromatographic analysis of benzodiazepines.[184,185]

Until very recently, most GC/MS assays for benzodiazepines involved either direct extraction and analysis of the underivatized drug,[186–188] or conversion of the benzodiazepine to its corresponding benzophenone by acid hydrolysis followed by extraction and GC/MS analysis (with or without derivatization).[183,189,190] Both of these general methods have significant limitations. The approach involving direct extraction and GC/MS analysis suffers from the fact that some benzodiazepines are thermally degraded when injected into a gas chromatograph;[191] moreover, many of the benzodiazepines are rapidly converted in the body to polar metabolites that are not suitable for GC/MS analysis without derivatization. Conversion of benzodiazepines to benzophenones gives a class of compounds that is generally easier to analyze by GC/MS, but since more than one benzodiazepine may be converted to the same benzophenone, this approach does not always permit identification of the specific benzodiazepine(s) that was present in the biological specimen. Furthermore, the triazolobenzodiazepines are not converted to benzophenones upon acid hydrolysis.[189]

To overcome these limitations, three recently published methods have included: (1) treatment with β-glucuronidase to hydrolyze conjugated metabolites, (2) solid-phase or liquid-liquid extraction, (3) derivatization, and (4) GC/MS analysis.[192–194]

1. Internal Standards

Many deuterium-labeled benzodiazepines and benzodiazepine metabolites are now sold by Sigma Chemical Co. (St. Louis, MO) and Radian (Austin, TX). The specific deuterated benzodiazepine compounds listed in the vendors' 1993 catalogs include: diazepam-d_5, desmethyldiazepam-d_5, oxazepam-d_5, chlordiazepoxide-d_5, alprazolam-d_5, α-hydroxyalprazolam-d_5, triazolam-d_5, α-hydroxytriazolam-d_4, flurazepam-d_5, lorazepam-d_4, nitrazepam-d_5, and temazepam-d_5. The deuterium-labeled isotopomers are generally the most satisfactory internal standards for analysis of specific benzodiazepines. However, many published procedures employ one nonlabeled benzodiazepine as an internal standard for measurement of another benzodiazepine. For example, medazepam has served as an internal standard for measurement of diazepam and N-desmethyldiazepam,[188] and triazolam as the internal standard for measurement of alprazolam.[186]

2. Extraction

Published procedures for removal of benzodiazepines from physiological specimens have included both liquid-liquid and solid-phase extractions. Diazepam can be efficiently extracted from alkalinized (pH 9) plasma with moderately polar solvents such as *tert*-butyl methyl ether[188] or toluene-ethyl acetate (4:1).[195] A more polar solvent (butyl acetate) was used for extracting 18 benzodiazepines and metabolites from plasma.[177] In a recently reported negative-ion chemical-ionization assay, Fitzgerald and co-workers hydrolyzed urine samples by treatment with β-glucuronidase (200 μl of 5000 U/ml in 1.0 M acetate buffer, pH 5.0) for 2 h at 70°C, adjusted the pH by addition of saturated sodium borate, and extracted the benzodiazepines and metabolites with toluene-hexane-isoamyl alcohol (78:20:2, v/v).[194] Extraction efficiencies ranged from 73 to 89%.

Solvents for liquid-liquid extraction of benzophenones formed by acid hydrolysis of benzodiazepines include dichloromethane:isopropanol:ethyl acetate (1:1:3)[183] and diethyl ether.[189]

Solid-phase extraction procedures are increasingly coming into favor for the GC/MS analysis of benzodiazepines. Five of the benzodiazepine analytes most often present in urine specimens (N-desmethyldiazepam, desalkylflurazepam, oxazepam, temazepam, and α-hydroxyalprazolam) were extracted with Bond Elut Certify columns after enzymatic hydrolysis and acidification with 1 M acetic acid.[193] The benzodiazepines were eluted from the columns with 3% ammonium hydroxide in ethyl acetate; recoveries ranged from 73 to 83%.

In other studies, Sep-Pak C_{18} cartridges were used to extract benzodiazepines from urine and plasma samples prior to wide-bore capillary gas chromatography. The samples were made alkaline and poured into activated C_{18} cartridges.[196] The cartridges were then washed with water and the benzodiazepines eluted with hexane-isopropanol (9:1). High recoveries were obtained for all benzodiazepines in urine. Recoveries greater than 90% could be obtained from plasma samples also, when an initial perchloric acid deproteinization step was included in the procedure. The same research group used Sep-Pak C_{18} cartridges to extract benzophenones resulting from acid hydrolysis of urine and blood specimens containing benzodiazepines.[190] After activation of the cartridge, the hydrolyzed sample was poured into the cartridge. The cartridge was then washed with water and eluted with chloroform. Excellent recoveries were reported for most benzophenones.

Six different silica-bonded phases were evaluated for solid-phase extraction of seven benzodiazepines from blood and urine prior to HPLC analysis.[197] The study included determination of the optimum percent methanol in water for washing and elution of the solid-phase sorbents.

Koves and Wells extracted triazolam from post-mortem blood by adsorption onto Amberlite XAD-2, and used dichloromethane to recover the adsorbed triazolam from the resin prior to GC/MS analysis.[187]

3. Derivatization

Acetylation,[183,198] silylation,[177,192,194,199,200] and alkylation[201] have all been used to improve the thermal stability and chromatographic behavior of benzodiazepines and benzophenones. Silylation is most often performed by heating the extract with BSTFA, but the more stable *tert*-butyldimethylsilyl derivatives were recently exploited in GC/MS assays designed to detect and measure the most important benzodiazepine analytes in urine.[192,193] The EI mass spectra of *tert*-butyldimethylsilyl derivatives are typically dominated by large $(M-57)^+$ ion peaks, which are generally well suited for selected ion monitoring. Limits of quantitation less than 10 ng/ml have been reported using this derivative.[192]

Acetylation also yields relatively stable derivatives and has been used in the qualitative identification of benzodiazepines and their acid hydrolysis products.[183,189] The EI mass spectra and retention indices for 64 acetylated hydrolysis products have been published.[183]

Joern described an interesting extractive alkylation procedure for determining diazolo- and triazolobenzodiazepines and their metabolites in urine.[201] After enzymatic hydrolysis, the hydrolysate is made basic and extracted with methylene chloride. Methyl iodide and tetrahexylammonium hydrogen sulfate are added to the extract. After evaporation of the organic phase, the methylated benzodiazepines are reconstituted in dimethylsulfoxide and partitioned between water and hexane. The derivatized benzodiazepines are recovered in the hexane layer in high yield, free of interfering co-extractants. The sensitivity of this method for determination of triazolobenzodiazepines in urine was shown to be substantially better than that achieved by earlier methods. However, in these studies hydroxyethylflurazepam and desalkylhydroxyflurazepam proved difficult to derivatize, and detection of chlordiazepoxide was variable.

4. GC/MS Conditions and Performance

GC/MS assays for benzodiazepines typically involve splitless injection and an oven temperature programmed to at least 300°C. The triazolobenzodiazepines and many of the diazolobenzodiazepines elute at oven temperatures near 300°C and give broad, tailed peaks unless the capillary column is of good quality and is well deactivated. Narrow-bore methylsilicone and phenylmethylsilicone capillary columns are used most often, although a more polar (HP-17) wide-bore capillary column has been reported to give better separation of benzodiazepine analytes.[196]

The method of ionization is an important consideration in developing a GC/MS assay for benzodiazepines. The major fragmentation processes occurring after electron ionization, as well as after positive- and negative-ion chemical ionization, have been identified for 24 benzodiazepines.[181] Most of the EI mass spectra contain abundant high mass ions and are suitable for either qualitative or quantitative analysis. However, because all pharmaceutically important benzodiazepines contain electron-capturing substituents, negative-ion chemical ionization is the technique of choice for quantitative GC/MS assays requiring high sensitivity.[186,187,194,202–204] The electron-capture negative-ion chemical ionization mass spectra of benzodiazepines are normally very simple, consisting of one or a few abundant high-mass ion clusters.[194] This feature is ideal for quantitative analysis, but may not provide sufficient specificity for analysis of forensic samples. A disadvantage of the negative-ion chemical ionization technique is that the spectra can be strongly affected by changes in the ion-source temperature and the presence of trace amounts of oxygen. The intensity of the molecular anion tends to decrease (sometimes dramatically) as the ion-source temperature is raised. Traces of oxygen in the ion source can also decrease the intensity of the molecular anion, with a corresponding increase in the M-1 ion.[205] Nevertheless, with adequate care, negative-ion chemical ionization can permit accurate measurement of benzodiazepines at concentrations well below 1 ng/ml. A comparison of ionization methods for determining alprazolam and triazolam concluded that negative-ion chemical ionization is 80 to 150 times more sensitive than either electron ionization or positive-ion chemical ionization.[186] In a separate study, the TMS derivatives of seven common benzodiazepines and metabolites gave negative ion chemical ionization signal-to-noise ratios from four to several thousand times greater than that obtained with either positive ion chemical ionization or electron ionization.[194]

Benzodiazepines are also efficiently ionized by proton-transfer positive-ion chemical ionization.[206] Because benzodiazepines contain basic functional groups, the protonated molecule (MH)$^+$ is usually the base peak when methane or other proton-donors are the reagent gases.

I.   Fentanyl and Its Analogs

Considering the high abuse potential presented by fentanyl and its analogs and the number of deaths attributed to abuse of these compounds,[207] it is surprising that so few reports of GC/MS assays for fentanyls have appeared in the forensic toxicology literature. No doubt that is partly a consequence of the difficulties associated with the detection and measurement of these very potent drugs. The major difficulty confronting the forensic toxicologist is developing an assay with sufficient sensitivity. An investigation of body fluids from overdose victims found that the mean fentanyl concentration in blood was 3 ng/ml and in urine 4 ng/ml.[208] A lethal dose of 3-methylfentanyl is reported to be as low as a few micrograms.[209] An additional difficulty is the number of different fentanyl-related compounds that may be encountered in forensic samples. The fact that fentanyl is relatively easy to synthesize, even in clandestine laboratories, has resulted in the illicit distribution of numerous congeners for which standards are seldom available.

Approximately 80% of a fentanyl dose is excreted in the urine within 72 h, primarily as metabolites resulting from N-dealkylation and hydrolysis of the amide linkage.[210] Consequently, analysis of urine specimens should include methods for detection of the metabolites.

1. Internal Standards

The synthesis of trideuterated fentanyl for use as an internal standard was described in 1981.[211] However, deuterated fentanyl has recently become commercially available (fentanyl-d_5, Radian, Austin, TX and Sigma, St. Louis, MO). Alfentanil served as an internal standard for a GC/MS analysis of fentanyl in blood, urine, and tissue samples from an overdose victim,[212] and a homolog of alfentanil was used for determination of fentanyl in whole blood by GC and GC/MS.[213] Non-fentanyl related compounds (oxycodone[214] and papaverine[215]) have also been used as internal standards for GC/MS analysis of fentanyl in blood and plasma, but there is little reason to recommend them for this purpose.

2. Extraction

Various common solvents were compared for extraction of fentanyl from aqueous solutions at pH 12.[214] A mixture of hexane and ethanol (19:1) gave the highest recovery (100%), but 1-chlorobutane yielded cleaner chromatograms and a recovery of 76%. Efficiencies for extraction of five fentanyl drugs from plasma were determined by addition of 1 ml of phosphate buffer (pH 9.2) to 2 ml of plasma and extraction with 5 ml of chloroform:n-heptane:isopropanol (50:33:17 v/v). The efficiencies ranged from 78 to 85%.

A back-extraction procedure was instrumental in achieving a limit of detection of 50 pg/ml of fentanyl in whole blood analyzed by GC/MS.[213] A 2-ml aliquot of blood was made basic by addition of 2 ml of saturated borate buffer (pH 9) and extracted with 1 ml of 1-chlorobutane. The organic extract was then extracted with 3 ml of 0.5 M sulfuric acid. The acid phase was washed with hexane and made basic by addition of 0.5 ml of saturated sodium hydroxide. Finally, the aqueous phase was extracted with 5 ml of 1-chlorobutane. All extractions were conducted with vapor-phase silylanized glassware.[216] The average recovery of fentanyl at a concentration of 1.0 ng/ml in blood was 84%.

3. Derivatization

Fentanyl has no acidic hydrogens requiring derivatization for the purpose of improving chromatographic behavior. However, an unusual derivatization introduces a fluorinated substituent in order to increase fentanyl's response with electron-capture detection.[217] The derivatization consists of treatment of fentanyl in acetonitrile with heptafluorobutyric anhydride in the presence of 4-(dimethylamino)pyridine at 75°C for 1 h. Most fentanyl-related compounds yield two derivatives corresponding to vinylogous heptafluorobutyramides (Figure 8). In an unpublished work (RLF), this derivatization method was investigated as a means of achieving high sensitivity with negative ion chemical ionization. The negative ion chemical ionization mass spectrum of the heptafluorobutyramide derivative of fentanyl (Figure 9) shows four abundant high-mass ions, corresponding to the molecular anion and sequential loss of three HF molecules. Unfortunately, in this preliminary study the sensitivity achieved by negative ion chemical ionization of this derivative was no better than that which could be achieved by positive ion chemical ionization of underivatized fentanyl.

An extractive alkylation procedure has been developed for determination of N-dealkyl metabolites of fentanyl and its analogs in urine.[210] Although this procedure was used in combination with gas chromatography/electron capture detection, it should be equally suitable for analysis by GC/MS. A 1-ml urine sample was made basic with 0.1 ml of 3 N sodium hydroxide and extracted with methylene chloride. The organic phase was washed twice with 1% sodium hydroxide and the fentanyl metabolites were back-extracted into 0.1 N hydrochloric acid. Methylene chloride was then added to the acid solution along with approximately 150 mg of solid sodium bicarbonate. After the mixture was cooled in an ice bath, 100 μl of pentafluoropropionic anhydride was added and the mixture was vortexed. The organic layer was transferred to an evaporation tube and concentrated by evaporation prior to GC analysis for the N-pentafluoropropionyl derivatives of the N-dealkyl metabolites. The origins of the major ions in the EI mass spectra of the derivatized metabolites of fentanyl and 3-methylfentanyl were proposed on the basis of high-resolution mass measurements.

4. GC/MS Conditions and Performance

Of the published GC/MS assays for fentanyl-related drugs in blood and plasma, the method developed by Watts and Caplan is the most fully

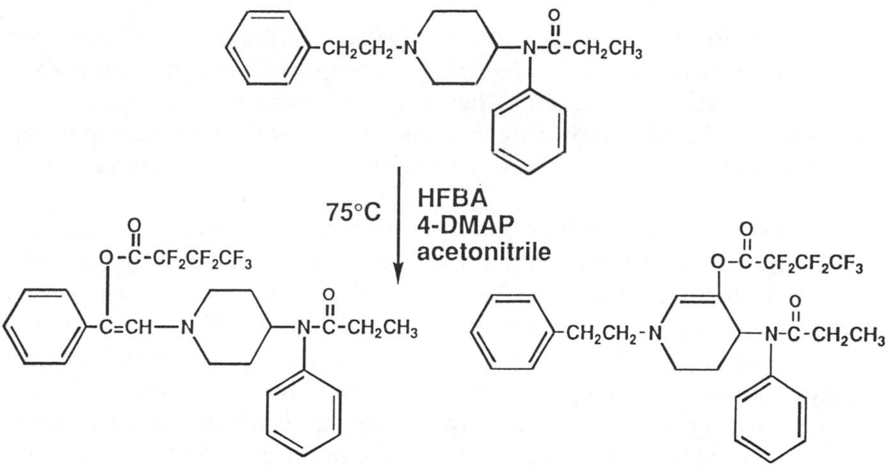

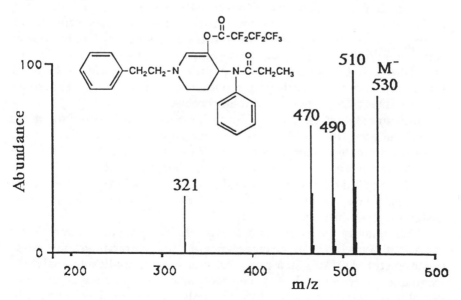

Figure 8.
Formation of fentanyl derivatives formed by treatment with heptafluorobutyric anhydride in the presence of 4-(dimethylamino)pyridine.

characterized in terms of sensitivity and reproducibility.[213] The method combines a selective extraction procedure, chromatographic separation on a 10-m × 0.1-mm i.d. capillary column coated with a 0.34-µm film of 5% phenyl methyl silicone, and electron ionization with selected ion monitoring. After injection, the GC column is held at 90°C for 1 min, then increased

Figure 9.
Negative ion chemical ionization mass spectrum of a heptafluorobutyrate derivative of fentanyl.

by 30°/min to 240°C, and by 5°/min to 280°C. Ultrahigh purity hydrogen as the carrier gas results in a fentanyl retention time of approximately 8 min. In validation experiments, the assay was linear for fentanyl concentrations from 50 pg/ml to 2.5 ng/ml; the between-run coefficient of variation for analysis of whole blood containing 0.4 ng/ml of fentanyl was 11.7%.

Methods employing chemical ionization have also achieved sub-ng/ml limits of quantitation. Following a direct extraction with ethyl acetate, extracts were injected into a 0.91-m × 2-mm i.d. glass column packed with 3% SE-30 and detected by positive ion chemical ionization with methane as the reagent gas. Linear responses were obtained for fentanyl plasma concentrations from 500 pg/ml to 100 ng/ml.[211] More recently, with ammonia as the reagent gas for CI analysis of fentanyl and N-dealkylfentanyl, the positive ion chemical ionization mass spectra of fentanyl and the pentafluoropropionyl derivative of N-dealkyl fentanyl each showed a single major peak corresponding to the protonated molecule (MH+).[218]

IV. SUMMARY AND FUTURE DIRECTIONS

The analysis of biological samples for drugs of abuse is, sadly, sure to continue to be an active area. The use and abuse of drugs is an age-old problem for humankind and promises to remain with us into the future. There will be changes, however. One certainty is that the drugs to be tested will change over time. While marijuana and cocaine have been abused for many years, other drugs leave or join the ranks of commonly abused substances. Some of these drugs are new and some are just the resurgence of old drugs. LSD is an example of a popular drug of the 1960s which fell into disfavor but has recently undergone a revival to the point that in some circles LSD is more widely used than cocaine.

Instrumentation associated with GC/MS analysis has become more reliable and versatile, and this trend promises to continue. The newest mass spectrometers offer greater sensitivity and stability, are easier to operate, and often cost less than earlier models. Much of the advance in ease of use is linked directly to rapid advancements in computers. Relatively inexpensive computers now have the capability of much larger systems of even just a few years ago.

Undoubtedly the most dramatic change in mass spectrometry over the past 10 years has been the emergence of liquid chromatography/mass spectrometry (LC/MS) as a powerful and widely applicable analytical technique. LC/MS has always been recognized as a potentially far more versatile technique than GC/MS, but only the recent development of atmospheric pressure interfaces has allowed LC/MS to rival GC/MS in terms of sensitivity.[219,220] Unfortunately, applications of LC/MS in forensic toxicology remain very limited, primarily because of the high cost of the instrumentation. Nevertheless, several recent publications have clearly demonstrated that not only can LC/MS be applied to the analysis of drugs and metabolites that are unsuited for GC/MS analysis,[221] but it has the potential of lowering the per-sample analytical cost by eliminating the need for derivatization and by allowing shorter analysis times.[222]

Tandem mass spectrometry (MS/MS) is another technique that clearly offers exciting capabilities for analysis of biological specimens, but its application in forensic toxicology has also been severely limited by the current cost and complexity of the instruments. Substantial efforts are being made to effect change in both cost and complexity. Most MS/MS analyses are currently being conducted on triple quadrupole instruments, in which an ion source first ionizes the sample; the first quadrupole selects an ion or a group of ions that then enter the second quadrupole (referred to as the collision cell), where a collision gas fed into the system causes the ions to collide and fragment further. The products of this collisionally induced dissociation (CID) are then monitored using the third quadrupole. MS/MS analyses have also been performed in ion trap mass spectrometers. After ionization within the trap, an excitation voltage is applied which causes the trapped ions to undergo collision-induced dissociations. The product ions are then ejected from the trap and detected by an electron multiplier.

MS/MS without chromatographic separation has been applied to the analysis of drugs in hair.[223,224] This technique can permit very rapid analysis. However, in most cases the improved sensitivity and specificity afforded by coupling MS/MS analysis to a chromatographic separation far outweigh the modest disadvantage of a slightly longer analysis time.

Automated sample preparation is also evolving rapidly. Commercially available robotic systems can process a sample aliquot from start to finish and produce a derivatized extract ready for injection into the GC/MS. The major limitations of current robotic systems are that they are costly and can process only one sample at a time. Active development in this area promises to provide substantially faster and more versatile systems in the near future.

REFERENCES

1. Klein, M., in *Forensic Mass Spectrometry*, Yinon, J., Ed., CRC Press, Boca Raton, FL, 1987, 51.
2. Liu, R. H., in *Forensic Mass Spectrometry*, Yinon, J., Ed., CRC Press, Boca Raton, FL, 1987, 1.
3. Yinon, J., *Mass Spectrom. Rev.*, 10, 179, 1991.
4. Trinh, V. and Vernay, A., *J. High Resolut. Chromatogr.*, 13, 162, 1990.
5. Chen, X.-H., Franke, J.-P., and de Zeeuw, R. A., *Forensic Sci. Rev.*, 4, 147, 1992.

6. Chen, X.-H., Wijsbeek, J., Franke, J.-P., and De Zeeuw, R. A., *J. Forensic Sci.*, 37, 61, 1992.
7. Ensing, K., Franke, J. P., Temmink, A., Chen, X.-H., and De Zeeuw, R. A., *J. Forensic Sci.*, 37, 460, 1992.
8. Lillsunde, P. and Korte, T., *J. Anal. Toxicol.*, 15, 71, 1991.
9. Platoff, G. E., Jr. and Gere, J. A., *Forensic Sci. Rev.*, 3, 117, 1991.
10. Harkey, M. R., in *Analytical Aspects of Drug Testing*, Deutsch, D. G., Ed., John Wiley & Sons, New York, 1989, 59.
11. Scheurer, J. and Moore, C. M., *J. Anal. Toxicol.*, 16, 264, 1992.
12. Moore, J. M., *Forensic Sci. Rev.*, 2, 80, 1990.
13. Knapp, D. R., *Methods Enzymol.*, 193, 314, 1990.
14. Knapp, D. R., *Handbook of Analytical Derivatization Reactions*, Wiley-Interscience, New York, 1979.
15. Blau, K. and Halket, J., Eds., *Handbook of Derivatives for Chromatography*, John Wiley & Sons, Chichester, 1993.
16. Grob, K., *Classical Split and Splitless Injections in Capillary GC*, Huethig Publishing, Ltd., 1988.
17. *Operating Hints for Split/Splitless Injectors*, Restek Corporation, Bellefonte, PA, 1992.
18. Harrison, A. G., *Chemical Ionization Mass Spectrometry*, CRC Press, Boca Raton, FL, 1992.
19. Inoue, T. and Seta, S., *Forensic Sci. Rev.*, 4, 89, 1992.
20. Moeller, M. R., *J. Chromatogr.*, 580, 125, 1992.
21. Cone, E. J., *Employment Testing*, 3, 439, 1989.
22. Harkey, M. R. and Henderson, G. L., in *Advances in Analytical Toxicology*, Vol. 2, Baselt, R. C., Ed., Year Book Medical Publishers, Chicago, 1989, 298.
23. Baumgartner, W. A., *Employment Testing*, 3, 442, 1989.
24. *Federal Register*, 53, 11970, 1988.
25. Garriott, J. C., Di Maio, V. J., and Rodriguez, R. G., *J. Forensic Sci.*, 31, 1274, 1986.
26. Huestis, M. A., Henningfield, J. E., and Cone, E. J., *J. Anal. Toxicol.*, 16, 283, 1992.
27. Foltz, R. L., in *Advances in Analytical Toxicology*, Vol. 1, Baselt, R. C., Ed., Biomedical Publ., Davis, CA, 1984, 125.
28. Harvey, D. J., *Mass Spectrom. Rev.*, 6, 135, 1987.
29. ElSohly, M. A., Little, T. L., and Stanford, D. F., *J. Anal. Toxicol.*, 16, 188, 1992.
30. ElSohly, M. A., Stanford, D. F., and Little, T. L., Jr., *J. Anal. Toxicol.*, 54, 1988.
31. Wu, A. H. B., Liu, N., Cho, Y.-J., Johnson, K. G., and Wong, S. S., *J. Anal. Toxicol.*, 17, 215, 1993.
32. Brunk, S. D., *J. Anal. Toxicol.*, 12, 290, 1988.
33. Foltz, R. L., McGinnis, K. M., and Chinn, D. M., *Biomed. Mass Spectrom.*, 10, 316, 1983.
34. Nelson, C. C., Fraser, M. D., Wilfahrt, J. K., and Foltz, R. L., *Ther. Drug Monitoring*, 15, 557, 1993.
35. Moeller, M. R., Doerr, G., and Warth, S., *J. Forensic Sci.*, 37, 969, 1992.
36. Rosenfeld, J., *Anal. Lett.*, 10, 917, 1977.
37. Ritchie, L. K., Caplan, Y. H., and Park, J., *J. Anal. Toxicol.*, 11, 205, 1987.
38. Johansson, E., Noren, K., Sjoevall, J., and Halldin, M. M., *Biomed. Chromatogr.*, 3, 35, 1989.
39. Lisi, A. M., Kazlauskas, R., and Trout, G. J., *J. Chromatogr.*, 617, 265, 1993.
40. Baker, T. S., Harry, J. V., Russell, J. W., and Myers, R. L., *J. Anal. Toxicol.*, 8, 255, 1984.
41. Clatworthy, A. J., Oon, M. C. H., Smith, R. N., and Whitehouse, M. J., *Forensic Sci. Int.*, 46, 219, 1990.
42. Wimbish, G. H. and Johnson, K. G., *J. Anal. Toxicol.*, 14, 292, 1990.
43. Dixit, V. and Dixit, V. M., *J. Chromatogr.*, 567, 81, 1991.
44. Nakamura, G. R., Stall, W. J., Masters, R. G., and Folen, V. A., *Anal. Chem.*, 37, 1492, 1985.

45. McCurdy, H. H., Lewellen, L. J., Callahan, L. S., and Childs, P. S., *J. Anal. Toxicol.*, 10, 175, 1986.
46. Paul, B. D., Mell, L. D., Jr., Mitchell, J. M., and McKinley, R. M., *J. Anal. Toxicol.*, 11, 1, 1987.
47. Kogan, M. J., Razi, J. A., Pierson, D. J., and Willson, N. J., *J. Forensic Sci.*, 31, 494, 1986.
48. Parry, R. C., Nolan, L., Shirey, R. E., Wachob, G. D., and Gisch, D. J., *J. Anal. Toxicol.*, 14, 39, 1990.
49. Parry, R. C. and Gisch, D. J., *LC-GC*, 7, 972, 1989.
50. Lemm, U., Tenczer, J., and Baudisch, H., *J. Chromatogr.*, 342, 393, 1985.
51. Law, B., Mason, P. A., Moffat, A. C., Gleadle, R. I., and King, L. J., *J. Pharm. Pharmacol.*, 36, 289, 1984.
52. Perez-Reyes, M. and Wall, M. E., *J. Clin. Pharmacol.*, 21, 178, 1981.
53. McBurney, L. J., Bobbie, B. A., and Sepp, L. A., *J. Anal. Toxicol.*, 10, 56, 1986.
54. Bronner, W. E. and Xu, A. S., *J. Chromatogr.*, 580, 63, 1992.
55. Whiting, J. D. and Manders, W. M., *Aviat. Space Environ. Med.*, 54, 1031, 1983.
56. Nakamura, G., Meeks, R., and Stall, W., *J. Forensic Sci.*, 35, 792, 1990.
57. Clouette, R., Jacob, M., Koteel, P., and Spain, M., *J. Anal. Toxicol.*, 17, 1, 1993.
58. Karlsson, L., Jonsson, J., Aberg, K., and Roos, C., *J. Anal. Toxicol.*, 7, 198, 1983.
59. Joern, W. A., *J. Anal. Toxicol.*, 11, 49, 1987.
60. Hughes, J. and Hoey, L. D., Hewlett-Packard, GC/MS 91–7, June 1991.
61. Levine, B. and Smith, M. L., *Forensic Sci. Rev.*, 2, 148, 1990.
62. Stewart, D. J., Inaba, T., Tang, B. K., and Kalow, W., *Life Sci.*, 20, 1557, 1977.
63. Baselt, R. C., *J. Chromatogr.*, 268, 502, 1983.
64. Isenschmid, D. S., Levine, B. S., and Caplan, Y. H., *J. Anal. Toxicol.*, 13, 250, 1989.
65. McCurdy, H. H., Callahan, L. S., and William, R. D., *J. Forensic Sci.*, 34, 858, 1989.
66. Dawling, S., Essex, E. G., Ward, N., and Widdop, B., *Ann. Clin. Biochem.*, 27, 478, 1990.
67. Brogan, W. C., Kemp, P. M., Bost, R. O., Glamann, D. B., Lange, R. A., and Hills, L. D., *J. Anal. Toxicol.*, 16, 152, 1992.
68. Cone, E. J. and Darwin, W. D., *J. Chromatogr.*, 580, 43, 1992.
69. Ambre, J., Ruo, T. I., Nelson, J., and Belknap, S., *J. Anal. Toxicol.*, 12, 301, 1988.
70. Ambre, J., *J. Anal. Toxicol.*, 9, 241, 1985.
71. Ambre, J. J., Ruo, T.-I., Smith, G. L., Backes, D., and Smith, C. M., *J. Anal. Toxicol.*, 6, 26, 1982.
72. Matsubara, K., Kagawa, M., and Fukui, Y., *Forensic Sci. Int.*, 26, 169, 1984.
73. Ambre, J., Fischman, M., and Ruo, T., *J. Anal. Toxicol.*, 8, 23, 1984.
74. Zhang, J. Y. and Foltz, R. L., *J. Anal. Toxicol.*, 14, 201, 1990.
75. Jindal, S. P. and Lutz, T., *J. Anal. Toxicol.*, 10, 150, 1986.
76. Cone, E. J. and Weddington, W. W., Jr., *J. Anal. Toxicol.*, 13, 65, 1989.
77. Rafla, F. K. and Epstein, R. L., *J. Anal. Toxicol.*, 3, 59, 1979.
78. Hime, G. W., Hearn, W. L., Rose, S., and Cofino, J., *J. Anal. Toxicol.*, 15, 241, 1991.
79. Wu, A. H. B., Onigbinde, T. A., Johnson, K. G., and Wimbish, G. H., *J. Anal. Toxicol.*, 16, 132, 1992.
80. Cone, E. J., Yousefnejad, D., Darwin, W. D., and Maguire, T., *J. Anal. Toxicol.*, 15, 250, 1991.
81. Janzen, K. E., *J. Forensic Sci.*, 36, 1224, 1991.
82. Mule, S. J. and Casella, G. A., *J. Anal. Toxicol.*, 12, 153, 1988.
83. Joern, W. A., *J. Anal. Toxicol.*, 11, 110, 1987.
84. Fletcher, S. M. and Hancock, V. S., *J. Chromatogr.*, 206, 193, 1981.
85. von Minden, D. L. and D'Amato, N. A., *Anal. Chem.*, 49, 1974, 1977.
86. Needleman, S. B., Goodin, K., and Severino, W., *J. Anal. Toxicol.*, 15, 179, 1991.
87. Gerlits, J., *J. Forensic Sci.*, 38, 1210, 1993.
88. Aderjan, R. E., Schmitt, G., Wu, M., and Meyer, C., *J. Anal. Toxicol.*, 17, 51, 1993.

89. Tebbett, I. R. and McCartney, Q. W., *Forensic Sci. Int.*, 39, 287, 1988.
90. Ellerbe, P., Tai, S. S.-C., Christensen, R. G., Espinosa-Leniz, R., Paule, R. C., Sander, L. C., Sniegoski, L. T., Welch, M. J., and White, E., *J. Anal. Toxicol.*, 16, 158, 1992.
91. Anderson, R. E. and Nixon, G. L., *J. Anal. Toxicol.*, 17, 432, 1993.
92. Matsubara, K., Maseda, C., and Fukui, Y., *Forensic Sci. Int.*, 26, 181, 1984.
93. Taylor, R. W. and Le, S. D., *Adv. Lab. Autom. Rob.*, 7, 567, 1991.
94. Schramm, W., Craig, P. A., and Smith, R. H., *Clin. Chem.*, 39, 481, 1993.
95. Abusada, G. M., Abukhaiaf, I. K., Alford, D. D., Vinzon-Bautista, I., Pramanik, A. K., Ansari, N. A., Manno, J. E., and Manno, B. R., *J. Anal. Toxicol.*, 17, 353, 1993.
96. Clark, G. D., Rosenzweig, I. B., Raisys, V. A., Callahan, C. M., Grant, T. M., and Streissguth, A. P., *J. Anal. Toxicol.*, 16, 261, 1992.
97. Lombardero, N., Casanova, O., Behnke, M., Eyler, F. D., and Bertholf, R. L., *Ann. Clin. Lab. Sci.*, 23, 385, 1993.
98. Isenschmid, D. S., Levine, B. S., and Caplan, Y. H., *J. Anal. Toxicol.*, 12, 242, 1988.
99. Taylor, R. W. and Le, S. D., *J. Anal. Toxicol.*, 15, 276, 1991.
100. Foltz, R. L., Fentiman, A. F., Jr., and Foltz, R. B., *GC/MS Assays for Abused Drugs in Body Fluids*, NIDA, Division of Research, Rockville, MD, 1980.
101. Dring, L. G. and Smith, R. L., *Biochem. J.*, 116, 425, 1970.
102. Beckett, A. H. and Rowland, M., *J. Pharm. Pharmacol.*, 17, 109S, 1965.
103. Autry, J. H., Notice to all DHHS/NIDA certified laboratories, Division of Applied Research, NIDA, Rockville, MD, 1992.
104. ElSohly, M. A., Stanford, D. F., Sherman, D., Shah, H., Bernot, D., and Turner, C. E., *J. Anal. Toxicol.*, 16, 109, 1992.
105. Gjerde, H., Hasvold, I., Pettersen, G., and Christophersen, A. S., *J. Anal. Toxicol.*, 17, 65, 1993.
106. Taylor, R. W., Le, S. D., Philip, S., and Jain, N. C., *J. Anal. Toxicol.*, 13, 293, 1989.
107. Mule, S. J. and Casella, G. A., *J. Anal. Toxicol.*, 12, 102, 1988.
108. Fitzgerald, R. L., Ramos, J. M., Jr., and Poklis, A., *J. Anal. Toxicol.*, 12, 255, 1988.
109. Gan, B. K., Baugh, D., Liu, R. H., and Walia, A. S., *J. Forensic Sci.*, 36, 1331, 1991.
110. Armbruster, D. A., Schwarzhoff, R. H., Hubster, E. C., and Liserio, M. K., *Clin. Chem.*, 39, 2137, 1993.
111. Lillsunde, P. and Korte, T., *Forensic Sci. Int.*, 49, 205, 1991.
112. Hornbeck, C. L. and Czarny, R. J., *J. Anal. Toxicol.*, 13, 144, 1989.
113. Ho, Y.-S., Liu, R. H., Nichols, A. W., and Kumar, S. D., *J. Forensic Sci.*, 35, 123, 1990.
114. Melgar, R. and Kelly, R. C., *J. Anal. Toxicol.*, 17, 399, 1993.
115. Czarny, R. J. and Hornbeck, C. L., *J. Anal. Toxicol.*, 13, 257, 1989.
116. Liu, J. H. and Ku, W. W., *Anal. Chem.*, 53, 2180, 1981.
117. Cody, J. T. and Schwarzhoff, R., *J. Anal. Toxicol.*, 17, 321, 1993.
118. Hughes, R. O., Bronner, W. E., and Smith, M. L., *J. Anal. Toxicol.*, 15, 256, 1991.
119. Cody, J. T., *J. Chromatogr.*, 580, 77, 1992.
120. DePace, A., Verebey, K., and ElSohly, M., *J. Forensic Sci.*, 35, 1431, 1990.
121. Marde, Y. and Ryhage, R., *Clin. Chem.*, 24, 1720, 1978.
122. Kojima, T., Une, I., and Yashiki, M., *Forensic Sci. Int.*, 21, 253, 1983.
123. Wu, A. H. B., Onigbinde, T. A., and Wong, S. S., *J. Anal. Toxicol.*, 16, 137, 1992.
124. Hayes, L. W., Krasselt, W. G., and Mueggler, P. A., *Clin. Chem.*, 33, 806, 1987.
125. Struempler, R. E., *J. Anal. Toxicol.*, 11, 97, 1987.
126. Zebelman, A. M., Troyer, B. L., Randall, G. L., and Batjer, J. D., *J. Anal. Toxicol.*, 11, 131, 1987.
127. ElSohly, H. N., ElSohly, M. A., and Stanford, D. F., *J. Anal. Toxicol.*, 14, 308, 1990.
128. Liu, R. H., *Forensic Sci. Rev.*, 4, 52, 1992.
129. Cone, E. J., Welch, P., Mitchell, J. M., and Paul, B. D., *J. Anal. Toxicol.*, 15, 1, 1991.
130. Goldberger, B. A., Darwin, W. D., Grant, T. M., Allen, A. C., Caplan, Y. H., and Cone, E. J., *Clin. Chem.*, 39, 670, 1993.

131. Tay, M. K., Lee, T. K., and Chiu, W. Y., *Ann. Acad. Med. (Singapore)*, 22, 11, 1993.
132. Chen, B. H., Taylor, E. H., and Pappas, A. A., *J. Anal. Toxicol.*, 14, 12, 1990.
133. Mitchell, J. M., Paul, B. D., Welch, P., and Cone, E. J., *J. Anal. Toxicol.*, 15, 49, 1991.
134. Yeh, S. Y., Gorodetzky, C. W., and McQuinn, R. L., *J. Pharmacol. Exp. Ther.*, 196, 249, 1976.
135. Combie, J., Blake, J. W., Nugent, T. E., and Tobin, T., *Clin. Chem.*, 28, 83, 1982.
136. Jennison, T. A., Wozniak, E., Nelson, G., and Urry, F. M., *J. Anal. Toxicol.*, 17, 208, 1993.
137. Zezulak, M., Snyder, J. J., and Needleman, S. B., *J. Forensic Sci.*, 38, 1275, 1993.
138. Paul, B. D., Mell, L. D., Mitchell, J. M., Irving, J., and Novak, A. J., *J. Anal. Toxicol.*, 9, 222, 1985.
139. Bermejo, A. M., Fernandez, R. P., Lopez-Rivadulla, M., Cruz, A., Chiarotti, M., Fucci, N., and Marsilli, R., *J. Anal. Toxicol.*, 16, 372, 1992.
140. Chen, X.-H., Hommerson, A. L. C., Zwepfenning, P. G. M., Franke, J.-P., Harmen-Boverhof, C. W., Ensing, K., and de Zeeuw, R. A., *J. Forensic Sci.*, 38, 668, 1993.
141. Grinstead, G. F., *J. Anal. Toxicol.*, 15, 293, 1991.
142. Bowie, L. J. and Kirkpatrick, P. B., *J. Anal. Toxicol.*, 13, 326, 1989.
143. Fuller, D. C. and Anderson, W. H., *J. Anal. Toxicol.*, 16, 315, 1992.
144. Cone, E. J., Darwin, W. D., and Buchwald, W. F., *J. Chromatogr.*, 275, 307, 1983.
145. Paul, B. D., Mitchell, J. M., and Mell, L. D., Jr., *J. Anal. Toxicol.*, 13, 2, 1989.
146. Schuberth, J. and Schuberth, J., *J. Chromatogr.*, 490, 444, 1989.
147. Liu, R. H., Low, I. A., Smith, F. P., Piotrowski, E. G., and Hsu, A., *Org. Mass Spectrom.*, 20, 511, 1985.
148. Phillips, W. H., Jr., Ota, K., and Wade, N. A., *J. Anal. Toxicol.*, 13, 268, 1989.
149. Pocci, R., Dixit, V., and Dixit, V. M., *J. Anal. Toxicol.*, 16, 45, 1992.
150. Mule, S. J. and Casella, G. A., *J. Anal. Toxicol.*, 13, 13, 1989.
151. Benschenkroun, Y., Ribon, B., Desage, M., and Brazier, J. L., *J. Chromatogr.*, 420, 287, 1987.
152. Hyde, P. M., *J. Anal. Toxicol.*, 9, 269, 1985.
153. Gambaro, V., Mariani, R., and Marozzi, E., *J. Anal. Toxicol.*, 6, 321, 1982.
154. Barbour, A. D., *J. Anal. Toxicol.*, 15, 214, 1991.
155. Maurer, H. H., *J. Chromatogr.*, 530, 307, 1990.
156. Brzezinka, H., Bold, P., and Budzikiewicz, H., *Biol. Mass Spectrom.*, 22, 346, 1993.
157. Domino, E. F. and Wilson, A. E., *Clin. Pharmacol. Ther.*, 22, 421, 1977.
158. Stevenson, C. C., Cibull, D. L., Platoff, G. E., Jr., Bush, D. M., and Gere, J. A., *J. Anal. Toxicol.*, 16, 337, 1992.
159. ElSohly, M. A., Little, T. L., Jr., Mitchell, J. M., Paul, B. D., Mell, L. D., Jr., and Irving, J., *J. Anal. Toxicol.*, 12, 180, 1988.
160. Kelly, R. C. and Christmore, D. S., *J. Forensic Sci.*, 27, 827, 1982.
161. Lin, D. C. K., Fentiman, A. F., Jr., Foltz, R. L., Forney, R. D., Jr., and Sunshine, I., *Biomed. Mass Spectrom.*, 2, 206, 1975.
162. Kidwell, D. A., *J. Forensic Sci.*, 38, 272, 1993.
163. *Drug Detection Report*, 5, 1992.
164. Seligmann, J., Mason, M., Annin, P., Marszalek, D., and Wolfberg, A., *Newsweek*, p. 66, 7, 1992.
165. Paul, B. D., Mitchell, J. M., Burbage, R., Moy, M., and Sroka, R., *J. Chromatogr.*, 529, 103, 1990.
166. Nelson, C. C. and Foltz, R. L., *J. Chromatogr.*, 580, 97, 1992.
167. Foltz, R. B. and Foltz, R. L., in *Advances in Analytical Toxicology*, Vol. 2, Baselt, R. C., Ed., Year Book Medical Publishers, Chicago, IL, 1989, 140.
168. Francom, P., Lim, H. K., Andrenyak, D., Jones, R. T., and Foltz, R. L., *J. Anal. Toxicol.*, 12, 1, 1988.
169. Lim, H. K., Andrenyak, D., Francom, P., Jones, R. T., and Foltz, R. L., *Anal. Chem.*, 60, 1420, 1988.

170. Nelson, C. C. and Foltz, R. L., *Anal. Chem.*, 64, 1578, 1992.
171. Papac, D. I. and Foltz, R. L., *J. Anal. Toxicol.*, 14, 189, 1990.
172. Kidwell, D. A. and Vincenz, M. S., Naval Research Laboratory, NRL/MR/6177–92–7180, December 28, 1992.
173. Sun, J., *Am. Clin. Lab.*, 8, 24, 1989.
174. Bukowski, N. and Eaton, A. N., *Rapid Commun. Mass Spectrom.*, 7, 106, 1993.
175. *Analytical Profiles of the Benzodiazepines*, CND Analytical, Auburn, AL, 1989.
176. Zamecnik, J., Ethier, J., and Neville, G., *Can. Soc. Forensic Sci. J.*, 22, 233, 1989.
177. Drouet-Coassolo, C., Aubert, C., Coassolo, P., and Cano, J. P., *J. Chromatogr.*, 487, 295, 1989.
178. Ardrey, R. E., Allan, A. R., Bal, T. S., Joyce, J. R., and Moffat, A. C., Eds., *Pharmaceutical Mass Spectra*, The Pharmaceutical Press, London, 1985.
179. Weston, S. I., Japp, M., Partridge, J., and Osselton, M. D., *J. Chromatogr.*, 538, 277, 1991.
180. Japp, M., Garthwaite, K., Geeson, A. V., and Osselton, M. D., *J. Chromatogr.*, 439, 317, 1988.
181. Hattori, H., Suzuki, O., Sato, K., Mizutani, Y., and Yamada, T., *Forensic Sci. Int.*, 35, 165, 1987.
182. Cortes, E. C., Martinez, R., Ugalde, M., and Maldonado, N., *Org. Mass Spectrom.*, 26, 113, 1991.
183. Maurer, H. and Pfleger, K., *J. Chromatogr.*, 422, 85, 1987.
184. Sioufi, A. and Dubois, P., *J. Chromatogr., Biomed. Appl.*, 531, 459, 1990.
185. Jones, G. R. and Singer, P. P., in *Analytical Toxicology*, Vol. 2, Baselt, R. C., Ed., Year Book Medical Publishers, Chicago, 1989, 1.
186. Javaid, J. I. and Liskevych, U., *Biomed. Mass Spectrom.*, 13, 129, 1986.
187. Koves, G. and Wells, J., *J. Anal. Toxicol.*, 10, 241, 1986.
188. Kudo, K., Nagata, T., Kimura, K., Imamura, T., and Noda, M., *J. Chromatogr.*, 431, 353, 1988.
189. Jones, C. E., Wians, F. H., Jr., Martinez, L. A., and Merritt, G. J., *Clin. Chem.*, 35, 1394, 1989.
190. Seno, H., Suzuki, O., Kumazawa, T., and Hattori, H., *J. Anal. Toxicol.*, 15, 21, 1991.
191. Joyce, J. R., Bal, T. S., Ardrey, R. E., Stevens, H. M., and Moffat, A. C., *Biomed. Mass Spectrom.*, 11, 284, 1984.
192. Dickson, P. H., Markus, W., McKernan, J., and Nipper, H. C., *J. Anal. Toxicol.*, 16, 67, 1992.
193. West, R. E. and Ritz, D. P., *J. Anal. Toxicol.*, 17, 114, 1993.
194. Fitzgerald, R. L., Rexin, D. A., and Herold, D. A., *J. Anal. Toxicol.*, 17, 342, 1993.
195. Duthel, J. M., Constant, H., Vallon, J. J., Rochet, T., and Miachon, S., *J. Chromatogr.*, 579, 85, 1992.
196. Suzuki, O., Seno, H., and Kumazawa, T., *J. Forensic, Sci.*, 33, 1249, 1988.
197. Casas, M., Berrueta, L. A., Gallo, B., and Vicente, F., *J. Pharmaceut. Biomed. Anal.*, 11, 277, 1993.
198. Brooks, K. E. and Smith, N. B., *Clin. Chem.*, 35, 2100, 1989.
199. Mule, S. J. and Casella, G. A., *J. Anal. Toxicol.*, 13, 179, 1989.
200. Fraser, A. D., Bryan, W., and Isner, A. F., *J. Anal. Toxicol.*, 15, 25, 1991.
201. Joern, W. A., *J. Anal. Toxicol.*, 16, 363, 1992.
202. Koves, E. M. and Yen, B., *J. Anal. Toxicol.*, 13, 69, 1989.
203. Garland, W. A. and Min, B. H., *J. Chromatogr.*, 172, 279, 1979.
204. Miwa, B. J., Garland, W. A., and Blumenthal, P., *Anal. Chem.*, 53, 793, 1981.
205. Garland, W. A. and Miwa, B. J., *Biomed. Mass Spectrom.*, 10, 126, 1983.
206. Miwa, B. H. and Garland, W. A., *J. Chromatogr.*, 139, 121, 1977.
207. Henderson, G. L., *J. Forensic Sci.*, 33, 569, 1988.
208. Henderson, G. L., *J. Forensic Sci.*, 36, 422, 1991.

209. Henderson, G. L., *Proc. West. Pharmacol. Soc.*, 26, 287, 1983.
210. Hammargren, W. R. and Henderson, G. L., *J. Anal. Toxicol.*, 12, 183, 1988.
211. Lin, S. N., Wang, T. P. F., Caprioli, R. M., and Mo, B. P. N., *J. Pharm. Sci.*, 70, 1276, 1981.
212. Matejczyk, R. J., *J. Anal. Toxicol.*, 12, 236, 1988.
213. Watts, V. and Caplan, Y., *J. Anal. Toxicol.*, 12, 246, 1988.
214. Pare, E. M., Monforte, J. R., Gault, R., and Mirchandani, H., *J. Anal. Toxicol.*, 11, 272, 1987.
215. van Rooy, H. H., Vermeulen, N. P. E., and Bovill, J. G., *J. Chromatogr.*, 223, 85, 1981.
216. Fenimore, D. C., Davis, C. M., Whitford, J. H., and Harrington, C. A., *Anal. Chem.*, 48, 2289, 1976.
217. Moore, J. M., Allen, A. C., Cooper, D. A., and Carr, S. M., *Anal. Chem.*, 58, 1656, 1986.
218. Sakashita, C. O., Peat, J. J., Peat, M. A., and Foltz, R. L., *Proceedings of the 33rd Annual Conference on Mass Spectrometry and Allied Topics*, 1985, 462.
219. Bruins, A. P., *Mass Spectrom. Rev.*, 10, 53, 1991.
220. Henion, J. and Lee, E., in *Practical Spectroscopy*, Vol. 8, Brame, E. G., Jr., Ed., Marcel Dekker, New York, 1990, 469.
221. Rule, G., McLaughlin, L. G., and Henion, J., *Anal. Chem.*, 65, 857A, 1993.
222. Covey, T. R., Lee, E. D., Bruins, A. P., and Henion, J. D., *Anal. Chem.*, 58, 1451A, 1986.
223. Pelli, B., Traldi, P., Tagliaro, F., Lubli, G., and Marigo, M., *Biomed. Environ. Mass Spectrom.*, 14, 63, 1987.
224. Martz, R. M., *Crime Lab. Dig.*, 15, 67, 1988.

Chapter 2

FORENSIC DRUG TESTING BY MASS SPECTROMETRIC ANALYSIS OF HAIR

Werner A. Baumgartner, Chen-Chih Cheng, Thomas D. Donahue, Gene F. Hayes, Virginia A. Hill, and Henry Scholtz

CONTENTS

I. FORENSIC ISSUES OF HAIR ANALYSIS

A. Probative Advantages of Hair

The detection and quantitative measurement of drugs in the hair of human subjects, first reported in 1979,[1] has proven to be a powerful clinical and forensic tool for the investigation of the use or abuse of licit and illicit drugs. One important feature of hair analysis is its long-term information on an individual's drug use in contrast to the short-term information provided by urinalysis.[2] Head hair grows at approximately 1.3 cm/month. Consequently, by sampling the segment of hair corresponding to a particular time frame, hair analysis can uncover drug use from a week to years prior to collection of the specimen.

Although hair and urine analysis are complementary with respect to time, in its probative aspects hair analysis has distinct advantages over urinalysis. For example, hair analysis cannot be thwarted by temporary abstention from drug use or adulteration of the specimen (this is a particularly important feature in pre-employment testing). Furthermore, the authenticity of the specimen can be guaranteed by closely supervised noninvasive collection procedures. In case of a challenge to the results of the first specimen, or a broken chain of custody, a second specimen can be collected and matched to the first one by its physical appearance, microscopic examination, or DNA analyses.[3] Entrapped analytes are protected by the stable matrix of hair. Consequently, hair can be stored almost indefinitely without refrigeration.[2,4] Hair analysis provides information on both the temporal pattern and the severity of drug use.[2,5-7] Its wide window of detection and permanent record of drug use make hair analysis far more effective than unannounced urinalysis for identification of even infrequent drug use.[7,8]

The advantages of hair analysis have been recognized in Congressional testimony,[9] by William Bennett in his National Drug Control Strategy,[10] and by the National Institute of Justice.[11] For epidemiological purposes, hair analysis has been recommended by the U.S. General Accounting Office[12] and is currently being used by the United Nations Interregional

Crime and Justice Research Institute.[13] Hair analysis has also attracted the attention of professional groups charged with meeting the challenge of drug abuse. For instance, in the clinical and criminal justice arenas, an indication of the severity of drug use through hair analysis greatly facilitates appropriate treatment or custody referrals.[7,8] Furthermore, efficacy of treatment (as measured by recidivism) as well as recovery status (i.e., whether drug use is decreasing, increasing, or remaining constant) can be readily ascertained by segmental hair analysis.[14] The wide window of drug detection in hair can provide an excellent indication of past drug use. Thus, it offers unprecedented information on prenatal drug exposure and on the epidemiology of drug use.[5,6,14,15] Hair analysis has also been very useful in assisting with challenges to positive urine test results such as claims of ingestion of poppy seed, spiked food or drink, or passive drug exposure.[9]

In spite of these unique advantages of hair analysis for clinical and epidemiological investigations, its main use to date has been in the forensic arena and in workplace drug testing where forensic procedures are also required. This chapter focuses, therefore, on forensic methods, as distinct from clinical or criminal justice methods of testing. In addition to mass spectrometric procedures, this report also describes criteria and methods for distinguishing between drugs incorporated into hair as a result of external contamination (exogenous drugs) and those incorporated into hair by either passive or active ingestion of drugs (endogenous drugs). A likely mechanism for the incorporation of endogenous drug into the interior structures of the hair shaft is via the circulatory system feeding the hair root (Figure 1).

B. Endogenous Evidentiary False Positives

One of the most important probative issues of both hair and urine drug testing is the avoidance of false positives due to passive exposure to drugs present in the environment. Such positives are described herein as evidentiary false positives to distinguish these from technical false positives which may be caused by errors in the analysis of hair or urine specimens.

Evidentiary false positives may occur either by endogenous or exogenous processes. Endogenous evidentiary false positives occur from drugs which enter blood as a result of passive drug exposure, whereas exogenous evidentiary false positives occur through direct contact of the drug with the urine or hair specimen. An example of the former is the passive inhalation of cocaine vapor from the environment which then enters urine or deep-lying interior structures of the hair via blood. These small quantities of endogenously absorbed drug cannot be removed by washing of

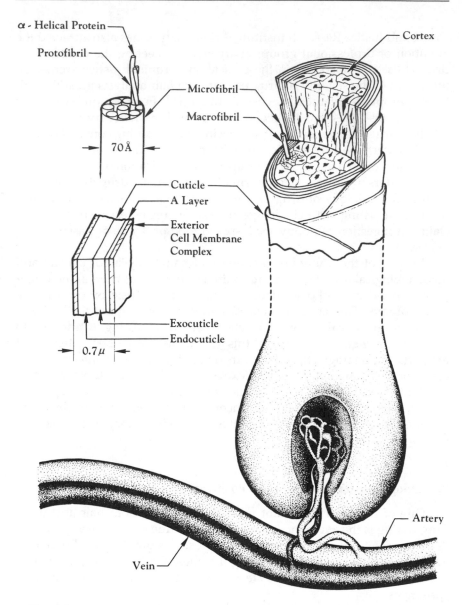

Figure 1.
Transfer of drug from the circulatory system to the hair follicle and its subsequent encapsulation in keratin fibers of hair shaft.

the hair specimen (see below). This is not a problem for hair analysis, since hair accumulates drugs in amounts approximately proportional to the ingested dose. With such small passively absorbed amounts of drug, drug concentrations in hair fall well below the endogenous cutoff levels set for

TABLE 1.

Cutoff Values Used in Drug Testing

Substance	Urine (ng/ml) Immunoassay	GC/MS	Hair (ng/g) Immunoassay	GC/MS
Marijuana metabolites	100		5	
Δ9-Carboxy-THC		15		0.1
Cocaine equivalents	300		500	
Cocaine/benzoylecgonine		150		500[a]
Opiates	300		500	
Morphine		300		300
Codeine		300		300
Phencyclidine (PCP)	25	25	300	300
Amphetamines	1000		500	
Amphetamine		500		300
Methamphetamine		500		300

[a]   Total of cocaine plus benzoylecgonine.

hair analysis (see below). Unfortunately, as the experience with the ingestion of the small amounts of morphine present in poppy seeds has shown, the endogenous cutoff levels of urinalysis do not provide the same margin of safety as the cutoff levels for hair analysis.

When the endogenous cutoff levels for hair are expressed in units comparable to urine (i.e., per gram of hair and per milliliter of urine), then, except for marijuana, the cutoff levels for hair and urine have very similar values (Table 1). This similarity, however, must not be interpreted as affording equal protection against endogenous evidentiary false positives. The only aspect that the cutoff values reveal about hair and urine testing is the need for greater sensitivity for mass spectrometric confirmations of positive hair samples. One reason for this is the small amount of hair generally available for employment testing (i.e., 20 to 30 mg). This is approximately a thousand times less than what is available for urinalysis (i.e., 30 ml). Hence hair analysis requires the application of ultrasensitive GC/MS/MS or ion trap procedures.

Quite a different approach than a simple comparison of cutoff levels is needed for an evaluation of the relative effectiveness of the endogenous cutoff levels used in urine and hair analysis to guard against endogenous evidentiary false positives. The relevant parameter in this regard is the ratio of the maximum attainable endogenous drug concentration in the hair or urine specimen to the drug concentration at the cutoff level (the risk ratio). The maximum concentration of the numerator should refer, of course, to the maximum concentration achieved as a result of drug use and not simply spiking of the sample.

If the risk ratio approaches unity because of the adoption of high cutoff levels, then the probability of the occurrence of an endogenous evidentiary false positive approaches zero. Of course, the effectiveness of identifying drug users also approaches zero under these conditions. As this ratio increases by lowering the cutoff level, the detection efficiency of the test increases, but so also does the probability of endogenous evidentiary false positives. The probability of the latter, however, is not a linear but rather an S-shaped function of the cutoff level. Thus, by the judicious choice of cutoff levels, the effectiveness of detecting drug users is balanced against the safety of a particular test.

With hair, the above-defined risk ratio is approximately 200 for cocaine and 20 for opiates, methamphetamine, marijuana, and phencyclidine. The peak endogenous drug concentrations of thoroughly washed hair for cocaine are approximately 1000 ng/10 mg hair; for opiates, phencyclidine, and methamphetamine, 100 ng/10 mg hair, and for marijuana, 10 pg/10 mg of hair.[2]

With urine, the risk ratios are much greater than with hair; thus, urinalysis poses much greater risks with respect to endogenous evidentiary false positives than hair. This situation can be readily demonstrated by several well-established examples. For instance, Selavka[16] has shown that the ingestion of the small quantities of opiates present in poppy seed can produce urine values over 6000 ng/ml, i.e., psychotropically insignificant amounts of ingested drugs can exceed the risk ratio of 20 used by hair analysis for maximum (not negligible) drug ingestion. Similarly, Baselt has shown that the passive ingestion of small amounts of cocaine may produce evidentiary false positive results.[17] Except for marijuana, the same appears likely for the other drugs. Marijuana is an exception in this regard because various reasons led to the adoption of a safer urine cutoff level (50 ng/ml).

Compared to hair analysis, the greater risk factor of urinalysis and its lower effectiveness in detecting drug users is largely due to the transient presence of drugs in urine. This makes the urine test around the time of maximum drug excretion ultrasensitive, i.e., too unsafe, and during the terminal phase of drug excretion, essentially nonsensitive. During this latter phase, drug users, so to speak, fall through the cracks (negative pharmacokinetic periods) of the urine test.

Hair analysis, on the other hand, does not suffer from these difficulties because the test is time-insensitive — i.e., the hair strand carries within its structure a permanent record of both the severity as well as the chronicity of drug use. It is this feature which bestows on hair analysis its greater safety and detection efficiency. Only when drug use is very frequent does urinalysis approach the detection efficiency of hair analysis,[8] but this occurs only under conditions of unannounced testing and when evasive maneuvers cannot be applied, i.e., under nonpre-employment testing conditions.

The greater safety of hair analysis with respect to endogenous passive drug exposure has been demonstrated in a study in which test subjects chronically ingested poppy seed confectionery for 1 month.[2] The opiate concentrations in hair, but not in urine, were well below the cutoff level. Randomly collected urine specimens, however, reached levels of 500 ng/ml.

This laboratory's experience in three U.S. Navy court martials and in other cases of challenged positive urine tests has shown that poppy seeds are by no means the only cause of evidentiary false positive urinalysis results. Selavka[16] and Baselt[17] have reported that individuals who are chronically exposed to a drug-containing environment can yield such positives. In addition, there is always the possibility or the defense of subversive activities against an individual, e.g., spiking of food or drink. It would appear that it was challenges such as these that led the National Institutes on Drug Abuse (NIDA) to mandate that urinalysis not be a stand-alone test, but that a positive urinalysis result be supported by the clinical evaluation of a medical review officer. For example, in the case of a positive opiate result, the presence of needle marks is required to support a diagnosis of drug abuse. On the other hand, it has been suggested, because of the greater effectiveness of hair analysis for investigating challenges to positive urinalysis results, that the medical review officer should request hair analysis in such situations.[2,9,18]

C. Exogenous Evidentiary False Positives

Hair analysis and urinalysis face entirely different problems with respect to avoiding exogenous evidentiary false positives. In the case of urinalysis, exogenous false positives may occur by contamination of the specimen in the laboratory or at the collection site, or by subversive activities. Since laboratory personnel can be a source of specimen contamination, it is important that these be monitored with an evasion-proof test such as hair analysis. On the other hand, challenges to urine specimens on the basis of subversive activities (e.g., micro-injection of drugs through the plastic container, etc.) are countered by safeguarding of the urine specimen in secure locations and by strict chain-of-custody procedures. If any of these security measures are breached or not optimally documented, nothing can be done to correct the problem. Unlike hair, the urine specimen cannot be cleansed by washing, nor can a second essentially identical specimen be collected from the donor at a later time.

In the case of hair, exogenous evidentiary false positives due to direct deposition of drugs on or in hair can be avoided with a high degree of certainty. For one, drugs of abuse do not bind strongly to the surface of the hair.[14] Furthermore, hair, unlike the lungs and gastrointestinal tract, is highly resistant to penetration by exogenous drugs.[14] Consequently, drugs may be readily removed from hair by conventional hygienic practices — the analog of the cleansing of the interior milieu by urine excretion.

Of much greater importance, however, is the ability to effectively decontaminate hair in the laboratory by specially developed wash procedures (see Section II). This does not only involve extensive washing with solvents especially chosen to match the condition of the hair (porous or nonporous), but also the use of specially developed wash kinetic procedures.[14,19,20]

Furthermore, in contrast to urine, the entry of drugs into hair does not result in the formation of metabolites. Thus, the absence or presence of metabolites (in a certain ratio relative to the original drug, once again defined by cutoff values) provides powerful additional assurance against exogenous evidentiary false positives in the case of hair analysis. This laboratory has demonstrated the effectiveness of these procedures by the safe testing of over 250,000 hair specimens (1.25 million separate analyses).

II. SAMPLE PREPARATION TECHNIQUES

A. Surface Chemistry of Hair

The wash procedures and associated wash kinetic analysis employed in this laboratory are based on the fact that exogenous drugs of abuse penetrate hair only poorly and that these bind only weakly to hair proteins. The strong binding of heavy metals to the protein sulfhydryl groups present in hair, frequently cited against hair analysis, is quite a different and unrelated matter.

As far as external contamination by the presently studied drugs is concerned, hair consists of three distinct domains: the accessible, semi-accessible, and inaccessible domains. The existence of such domains is well established in the cosmetic industry[19] and has been subsequently discovered to hold also for drugs of abuse.[2,14,20]

The accessible domain consists mainly of the hair surface to which drugs present in the environment have ready access. This domain, but not the other two, is readily accessed by nonhair-swelling solvents such as dry ethanol and dry isopropanol. Consequently, the loosely surface-bound exogenous drug contaminants are readily removed by dry ethanol or dry isopropanol.

The semi-accessible domain consists of interior hair structures which are not readily accessible to environmental drug contaminants, e.g., those

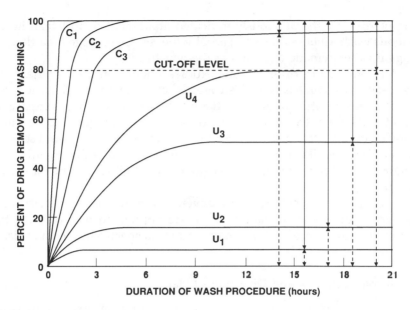

Figure 2.
Wash kinetic curves for the removal of drugs by phosphate buffer washes from the hair of drug users (U) and externally contaminated hair from nondrug-users (C). (↔) Amount of drug remaining in hair after attainment of plateau in the wash kinetics and subsequently released by digestion; (<– – –>) amount of drug removed by phosphate buffer washes.

in vapor form. However, this domain may be accessed to some degree by drugs in aqueous solution, possibly sweat. This domain, however, also contains drugs which have been deposited there by endogenous mechanisms, e.g., via the blood. Both endogenous and exogenous drugs can be completely removed from the semi-accessible domain by extensive and repeated washing with hair swelling solvents such as water, methanol, or water-containing ethanol or isopropanol solutions, but not with dry ethanol or isopropanol. Water, however, has been found to be the most effective wash solution in this regard. The depletion of the semi-accessible domain is signified by the attainment of a plateau in the wash kinetics most typically represented by curve U_2 in Figure 2 (see above).

The inaccessible domain is by far the largest domain. In strong thick hair it may represent more than 90% of the hair structure. This value, however, is significantly reduced if hair is damaged by the excessive application of hair treatments such as perming (permanent wave) or dyeing. Unless unrealistically severe contamination conditions are applied (elevated temperatures, time, and drug concentration), this inaccessible domain cannot be accessed by exogenous drugs in aqueous solutions. Consequently, endogenous drugs deposited in the inaccessible domain cannot be removed by extended water washes. This situation is characterized by the previously indicated attainment of a plateau in the aqueous wash kinetics (Figure 2) and by the release of large quantities of drugs upon enzymatic digestion of the hair fiber.

It appears from drug binding studies under conditions where the inaccessible domain has been destroyed by chemical means that the more effective sequestration of drugs in the inaccessible domain in comparison to that in the other two domains is not caused by stronger binding interactions with the protein structures of hair but rather by structural sequestration. The rope-like macroprotein structures as well as the organization of hair into cellular structures offer many possibilities for such a model (Figure 1).

Special wash procedures were developed on the basis of these well-established properties of hair.[14] These are applied to hair prior to the enzymatic digestion of the specimen and mass spectrometric analysis.

B. Decontamination and Digestion of Hair

Two types of wash procedures to free hair from exogenous drug contaminants have been adopted: (1) the extended wash procedure and (2) the truncated wash procedure. In the extended procedure, washes are continued until the wash kinetics attain a plateau. The duration and number of the washes are defined by the results obtained with a particular hair specimen. This procedure involves, therefore, the analysis of wash solutions while the wash procedure is being carried out. In the truncated procedure, the number of washes and their duration are fixed, and the results are related mathematically to the extended procedure by three wash kinetic criteria.[14]

The truncated wash procedure is used for mass production testing, and the extended wash procedure for the investigation of individualized forensic cases, or for samples which fall outside the truncated wash kinetic criteria, or in situations where the results of the first hair analysis with the truncated procedure are challenged ("safety net" samples). The kinetic analysis of the truncated wash procedure is described elsewhere,[14] since the present report deals only with individualized forensic testing.

Before commencing with the extended wash procedure, it is desirable to establish the porosity of the hair by methylene blue staining: 0.5% methylene blue, 5-min contact time followed by copious rinsing, and microscopic examination. The stain is readily taken up by porous hair but not by intact (nonporous) hair.

There are several causes for increased porosity of hair; the most frequent are cosmetic treatments such as perming, dyeing, and relaxing. In such cases, it is useful to separate the treated from the untreated portion of the hair by cutting the hair into appropriate segments. Other causes of increased permeability involved cases where the hair was maintained in an unfavorable environment, e.g., the 2-year-old remains of a murder victim, a 165-year-old hair specimen from the poet John Keats, and 600-year-old hair from a Peruvian mummy.[2,14]

An appropriate wash solvent is then chosen based on the porosity of the hair. Nonporous to slightly porous hair is washed with phosphate buffer (0.01 M, pH 5.6); porous hair with varying percentage mixtures of water and ethanol or isopropanol (e.g., 80/20 isopropanol/water). Isopropanol is preferred to ethanol to exclude the remote possibility of forming cocaethylene under *in vitro* conditions. Cocaethylene is a cocaine metabolite formed by the combined use of cocaine and ethanol.

As a preliminary to the phosphate washes, hair is also given a 15-min wash with dry isopropanol or dry ethanol. This step is performed: (1) to ensure that oils or greasy hair applications do not diminish the efficacy (i.e., alter the kinetics) of the wash procedure and (2) to remove loosely adhering drugs from the accessible domain — a necessary condition for the effective application of wash kinetic criteria (see below) for the distinction between endogenous and exogenous drugs.

In all cases, washing is performed at 37°C with shaking of the test tubes at 100 oscillations/min; 20 mg of hair in a 12- × 75-mm test tube are washed in 2 ml of wash solution. With the extended wash procedure, the phosphate solutions are withdrawn at convenient time intervals and replaced with fresh solution until the wash kinetics approach a plateau. Contact times of at least 30 min per interval should be allowed to ensure maximum hair swelling.

In the truncated wash procedure, hair is washed for 15 min in dry isopropanol and then three times for 30 min each in phosphate buffer. Aliquots of the wash solutions are analyzed by radioimmunoassay (RIA) for their drug content for the construction of wash kinetic curves. Mathematical relationships have been developed for the wash kinetic data, determined by RIA, and the drug content of washed hair as determined by the mass spectrometric analysis of the hair digest.

The drugs remaining in hair after a plateau has been reached in the extended wash procedure define the inaccessible domain of hair. The most effective method for releasing and measuring these residual drugs from the protein, but not the melanin, matrix is by the enzymatic digestion of the hair specimen. If solvent extraction procedures are used, complete extraction of analyte cannot be guaranteed, since extraction efficiency greatly depends on the physical properties of the hair, i.e., whether it is thin or thick, porous or nonporous, and upon the type and quantity of melanin present. The melanin may give rise to racial or hair color biases if solvent extraction procedures are used in place of digestion procedures. This would occur in cases where a major proportion of the analyte is sequestered in melanin granules. Such biases are not possible with the presently described hair digestion method, since the melanin fraction is removed and drugs are not released from the melanin fraction during the digestion process.

For mass spectrometric analyses of cocaine, benzoylecgonine, cocaethylene, phencyclidine, morphine, codeine, amphetamine, methamphetamine, and 11-nor-9-carboxy-9-tetrahydrocannabinol, a mild enzymatic digestion procedure was selected at pH 6.4, i.e., at a pH at which very little cocaine hydrolyzes to benzoylecgonine. Nevertheless, each hair digestion batch includes a hydrolysis control containing 100 ng of pure cocaine. This control is subsequently analyzed for its benzoylecgonine content. No special precautions need to be taken to measure the other metabolic analytes in hair.

The digestion of the hair specimen is performed as follows: 10 ml of 0.5 M Tris buffer, pH 6.4, to which is added 60 mg of dithiothreitol, 200 mg of sodium dodecyl sulfate, and 20 units of proteinase K. Add 2 ml of this solution per 20 mg of washed hair, the amount typically used for hair analysis. Place the mixture in a 37°C shaking water bath and shake 80 to 100 oscillations/min overnight (16 to 18 h). Remove from the water bath, mix, centrifuge, and remove the protein supernatant (leaving behind the melanin pellet) for further extraction, derivatization, and GC/MS/MS procedures. Because of the chemical complexity of the hair digest, special cleanup procedures harmless to GC equipment were developed for each of the analytes. These are described in detail in the experimental section on mass spectrometry.

It should be recognized that the chemical conditions which lead to the dissolution of one of the most resilient protein structures, hair, are highly deleterious to the most sensitive of proteins, antibodies. Consequently, the above described digest used for mass spectrometry is quite unsuitable for RIA analysis. For preliminary RIA screens for the identification of positive hair specimens, various solvent-based extraction procedures are available in the literature.[21–26]

C. Kinetic Analysis of Wash Data

The ratio of total amount of drug in the hair digest to the total amount of drug removed by all of the phosphate buffer washes with the extended wash procedure is an effective criterion for differentiating between drug use and external contamination. This criterion is called the Extended Safety Zone Ratio.

To illustrate the use of the Extended Safety Zone Ratio, consider the wash kinetics of several typical cases of external contamination. Curve C_1 in Figure 2 is an example where the external contamination was found mainly on the readily accessible surface domain, i.e., in this case the external contamination was almost completely and readily removed by washes with dry isopropanol; the small portion of remaining drugs was removed by the phosphate buffer washes, leaving insignificant quantities of drug (i.e., below the endogenous cutoff level) in the hair digest. The Extended Safety Zone Ratio for this case would be essentially zero.

In the case illustrated by curve C_2, the contamination of hair with an exogenous drug was applied under more severe conditions than in example C_1, i.e., a significant fraction of the exogenous drug penetrated the semi-accessible domain. Very little of this can be removed by dry isopropanol; however, essentially all of it is removed by washing with phosphate buffer. Once again, the Extended Safety Zone Ratio for this case is effectively zero.

Curve C_3 represents a case where external contamination was so severe as to cause some penetration into the highly resistant interior domain as measured by the drug content of the hair digest. Although the ratio of drugs in digest to drugs in phosphate buffer washes is quite small, the drug concentration in the digest may exceed the endogenous cutoff values listed in Table 1. However, the interpretation of such a result would still be one of "external contamination" on the basis of the wash kinetic cutoff criterion: the Extended Safety Zone cutoff. Its value for the extended wash procedure on the basis of empirical considerations has been set at 0.25.

To further illustrate the use of the Extended Safety Zone Ratio, let us approach the problem of distinguishing between drug use and external contamination from the other direction, i.e., from a study of the wash kinetics of hair from verified drug users. Curve U_1 in Figure 2 is a typical example of wash kinetics of a strong (thick) nonporous hair where drug use is clearly indicated by the wash kinetic results. Here very little or no drug was removed from the accessible and semi-accessible domains by phosphate buffer washes. However, upon digestion of the hair, a very large pulse of drug relative to the total amount in the phosphate buffer washes was released from the inaccessible domain. In other words, the Extended Safety Zone Ratio is a very large number relative to the most extreme contamination case illustrated by C_3.

Curve U_2 is typical of the wash kinetics of finer hair, where more endogenous drugs, and possibly exogenous drugs, are removed from the semi-accessible interior domain. However, the ratio of drugs in digest to total drugs in phosphate washes is still very high.

Curve U_3 is obtained with hair that has been rendered porous by treatments such as perming or dyeing. In extremely rare cases with very fine hair and excessively harsh hair treatment, hair may exhibit the wash kinetics represented by Curve U_4. Such cosmetic treatment reduces the size of the inaccessible domain. Considerably more endogenous and any exogenous drugs are removed by the phosphate washes. Correspondingly, the ratio of drugs in digest to drugs in phosphate washes is somewhat reduced. In the very rare case of U_4, the ratio may attain or exceed its cutoff value of 0.25. This ratio, of course, can be increased in such cases by the use of less harsh wash solutions such as 80:20 isopropanol:water mixture.

Hair samples producing Extended Safety Zone Ratios smaller than 0.25 are deemed to be contaminated on the basis of our extended wash kinetic criteria. Exceptions to this rule may occur when the metabolite criterion for drug use is applied (see below) and when no sign of hair damage is evident. Severe hair treatment or a heavily contaminated light user could present such findings. The opposite may also occur: borderline metabolite criteria may be over-ridden by highly definitive wash kinetic criteria. Highly definitive in this case means an Extended Safety Zone Ratio which is far above its cutoff value.

It follows from these considerations that in contrast to urine's dual reporting system of drug use or no drug use, a tripartite reporting system is available to hair analysis, i.e., (1) no drug use, (2) drug use, and (3) contaminated but no drug use. This tripartite method of reporting is useful in criminal justice-type situations where neither drug use nor being in the presence of drugs is permissible for probationers or parolees. An example of the latter would be a drug dealer who is not a drug user.

III. ELEMENTS OF MASS SPECTROMETRIC CONFIRMATION

A. Introduction

The interest in hair as an effective means of identifying both licit and illicit drug use has grown rapidly in recent years. Mass spectrometry as a confirmatory technique has become a strong ally to immunoassay screening procedures. With the various mass spectrometric methods available for the identification and confirmation of various drugs in human hair, it is timely to discuss the criteria for confirmation of presence. At the low levels found in the generally small hair samples, various data manipulations or alternate methods of approaching the analytical problem of confirmation have been employed to give acceptable forensic results. The problems experienced when dealing with nanogram and picogram levels in analysis are much more complex than when recording a mass spectrum of an ample supply of a reference standard. This review reflects the developing status of confirmation of drugs of abuse in hair through key published case histories.

Mass spectrometry is widely recognized as a reliable analytical tool because of its strengths in reproducibility, repeatability, specificity, and limit of detection. In a court of law, however, it is required to provide proof in a criminal case that is *beyond a reasonable doubt* or in a civil matter that *the preponderance of evidence* supports the conclusions. Because of the wide range in mass spectrometric techniques, interlaboratory validation in a newly emerging field such as hair analysis is often impractical. What has emerged over the last decade, however, is a set of criteria generally recognized as scientifically sound for obtaining reliable and quantitative results. To clarify this situation, a general set of guidelines has been formulated from the scientific literature to articulate the various approaches for confirmation irrespective of the ionization technique employed.

What lies ahead for MS in drug analysis of hair can be predicted with reasonable certainty. Low cost ion trap mass spectrometry will play an increasingly important role in its future developments, and structural elucidation will move from its classical focus on EI fragmentation to one where a greater burden is placed on chemical ionization and MS/MS procedures.

As early as 1978 it was concluded, on purely statistical grounds, that under electron ionization (EI) conditions, a minimum of three structurally significant ions are necessary to provide proof of presence.[27] A full mass spectral scan would, however, continue to provide the maximum and highest level of evidence. The intensity variation for ion abundance ratios was recommended to be within 5% when compared to a reference standard recorded under similar conditions.

However, because many of the softer ionization techniques do not cause extensive fragmentation, some EI confirmation criteria may not be directly applicable to CI conditions. Moreover, the experimental parameters governing the application of milder ionization techniques may not allow one to attain the same reproducibilities as observed under EI conditions.

The general principles governing the evolving criteria for confirmation of trace levels by mass spectrometry have been discussed recently[27] in terms of quality assurance controls, Good Laboratory Practices (GLPs), Good Measurement Practices (GMPs), Standard Operating Procedures (SOPs), Protocols for Specific Purposes (PSPs), recognized official methods status, tuning and calibration, mass spectral quality indices, the reality of variability, and limit of detection.

B. Gas Chromatography/Mass Spectrometry (GC/MS)

The synergistic support obtained in a mass spectral identification by the retention time of the eluting drug or metabolite is of obvious importance. For packed columns, a scan rate of 1 to 2 s/scan should be sufficient to enable reconstruction of the eluting profile to reflect accurately the original chromatographic resolving power. Faster scan rates can be used to increase the number of points to define the chromatographic elution pattern, whereas longer scan rates might well merge less concentrated components that occur with elutions close to a major component. Such distortions, if induced by either scanning too fast or too slow, severely hinder the acquisition of a representative mass spectrum from both the major and minor components for two reasons. First, at rapid scan rates, the ion statistics on data acquisition may begin to decline and also to affect signal-to-noise ratios. Second, at slower scan rates, the sample concentration in the source would not be constant during the data acquisition, and relative peak intensities might be perturbed.

With drug confirmations, the retention time correlation should normally be within the 2% error factor (as generally accepted for primary GC) compared to repeated injections of the sample or a reference standard. The accuracy of retention time correlations can be improved if the sample and standard are recorded within a relatively short period of time, usually only with a blank injection between them. However, there may be situations where compliance with the preferred 2% deviation guideline may not be attainable because of sample matrix problems.

In the case of capillary column GC/MS, the scan rate plays a more important role than in packed column analysis. Because the elution time frame of the average compound is only 2 to 4 s, a scan rate capable of acquiring several full scans is often necessary. In such cases, however, it should be recognized that the sample concentration in the source during a scan varies considerably. At scan rates of 1 s, ion statistics are usually not severely compromised.[27] Representative spectra, however, will be best acquired by combining all scans over the profile and taking the average. Capillary column conditions offer the main advantage of allowing a full mass spectral scan to be obtained on smaller quantities of material because the compound is concentrated in a very narrow band. Quite often this cannot be attained under packed column conditions, with the result that multiple ion detection (MID) approaches have to be applied.

At slow scan speeds relative to overall capillary elution times, the chromatographic reproduction of the retention time can only suffer from a slight perturbation of retention time correlation with a reference standard (at most 1 to 2 s). Such deviations are usually well within the 2% error factor expected from injections by the operator. Quantification using capillary GC/MS can only be satisfactorily achieved by employing MID descriptors whereby dwell times are selected to give at least ten data points over the elution profile.

To utilize the full potential of capillary GC/MS, temperature programming is often used to survey compounds over a wide range of polarity in a single analysis. Under such conditions, correlation with previously published data is impossible.

1. Methods of Ionization

Historically the fingerprinting capability of **electron impact (EI)** spectra, which benefits from the extensive number of fragments formed after ionization, has been the cornerstone of identification either by comparison with published data bases or structural identification via fragmentation patterns. In general, the reproducibility of the EI spectrum of a particular compound by magnetic instruments from one laboratory to another has been acceptable to the extent that direct comparisons were commonplace and reliable. However, with the advent of quadrupole mass spectrometers, this ability has been impaired. Because quadrupole instruments had the ability to mass discriminate during the tune-up procedures, direct comparisons with EI-generated spectra on magnetic sector instruments caused the correlation of relative abundance ratios from the previously accepted 5% to be extended to as much as 50%.

It is recognized that there often exists a variance between a previously recorded spectrum of a reference standard and an actual reported sample finding. For proof of presence, the recording of both sample and reference standard should be carried out on the same instrument on the same day under the same conditions. The widespread practice of using relative retention data and detector responses from two or more GC stationary phases of varying polarity has provided a high degree of certainty in the primary identification process. However, it must be recognized that there may be instances where two compounds may exhibit the same behavior, and hence the potential for misidentification can exist.

For EI confirmation purposes, the statistical need for three or more ions of structural significance observed in the correct relative abundance ratios has been clearly demonstrated.[27] Monitoring less than three ions can cause misidentification, and such analytical approaches may be considered screening techniques.

Positive ion chemical ionization (PICI) techniques to favor production of a molecule ion species for characterization purposes have been used for a number of reasons to resolve identification problems. First, the ability to determine the molecular weight of the compound under investigation has often been considered of paramount importance as the initial step in the identification process. Second, the use of chemical ionization may avoid the need to perform intensive sample cleanup by suppressing the interfering fragment ions. Third, the use of both methane and ammonia as reagent gases in separate experiments can often reveal the necessary structurally significant ions for confirmation, whatever the polarity of the analyte. However, the PICI approach may not always produce the optimum number of ions for confirmation.

Whereas electron capture (EC) is the least specific detector commonly used for pesticide analysis, the parallel between EC and **negative ion** chemical ionization (NICI) was quoted[28] as the main reason that this technique can be extrapolated effectively to drug analysis. The advantage of NICI lies in the possibility of greater sensitivity than PICI while providing complementary data. However, the same disproportionality plagues NICI as with EC. For compounds with a high-electron capture cross-section, this nonlinear response can start as low as 10 ng and saturate at 100 ng. However, the distinct advantage provided by such compounds is that their detection levels by NICI are in the picogram or lower range where quantitative support can reinforce GC measurements.

C. Mass Spectrometry/Mass Spectrometry (MS/MS)

1. Quantitative Aspects of Daughter Ions

The production of a protonated molecule ion for a particular drug via soft ionization methods, such as methane PICI, is generally preferred for primary identification purposes. Observance of molecular ions can be considered the most important criterion for identification, but the burden of proof of presence placed on a single ion species cannot be regarded as sufficient for confirmation. Although a single ion representing the molecule at the correct retention time on a packed column or a high resolution capillary might seem to have furnished sufficient evidence for confirmation, the need to prove structural dimensions still remains. With the introduction into commerce of MS/MS instruments, the possibility of improving the degree of specificity by soft ionization techniques has strongly emerged for practical drug confirmation.

Reliance on daughter ions has effectively replaced the former chromatographic retention data as a prime criterion for identification. This ability to unmask the chemical structure of the drug by daughter ions has recently found application in the field of anabolic steroids. In an MS/MS study of testosterone esters,[29] it was discovered that two generic daughter ions, m/z 97 and 109, were useful for identification of 4-en-3-one steroids. This discovery has opened up the possibility of a sophisticated reaction monitoring protocol for low level screening of such steroids in biological fluids. In addition to the two generic daughter ions, a daughter ion corresponding to the C-17 side chain was also present to assist with final structural elucidation.

The exploration of daughter ions for quantitative purposes was first investigated in the determination of ethyl carbamate in wines and spirits at the ppb level.[30] With the availability of a stable isotopically labeled ethyl carbamate reference standard, both confirmation and quantification were accomplished using methane chemical ionization MS/MS. The authors demonstrated that reliable quantitative data at low levels could be generated by isotope dilution techniques involving ratio measurements between daughter ions belonging to the sample and reference standard.

IV. CASE HISTORIES

A. Methamphetamine and Amphetamine

$$CH_3$$
$$CH_2CHNHCH_3$$

$$CH_3$$
$$CH_2CHNH_2$$

METHAMPHETAMINE   AMPHETAMINE

Suzuki et al.[31] described a GC/MS confirmation procedure in which hair samples, once extracted, were derivatized with trifluoroacetic anhydride (TFA) using N-methylbenzylamine as an internal standard. A packed column GC in conjunction with methane chemical ionization was used to enhance the production of the protonated molecular ions for all three compounds — m/z 246 for methamphetamine, m/z 232 for amphetamine, and m/z 218 for the internal standard. Under methane CI conditions, two or more structurally significant fragment ions were available for confirmation of structure. Quantification, however, was carried out using the protonated molecular ions in a multiple ion detection (MID) mode. The detection limit claimed, 100 pg injected on column, allowed a single strand of hair to be used for typical analysis. The calibration curves indicated linearity over several orders of magnitude for both drugs.

On the same theme, Nakahara et al. advanced this derivatization approach using TFA by employing stable isotopes, the d_4 analogs of methamphetamine and amphetamine, to improve the quantification step.[32,33] Chromatographic separation was done with a megabore column. These authors chose the fragment ions under EI, namely the base peaks, for the necessary isotope ratio measurements, m/z 154/158 for metham-

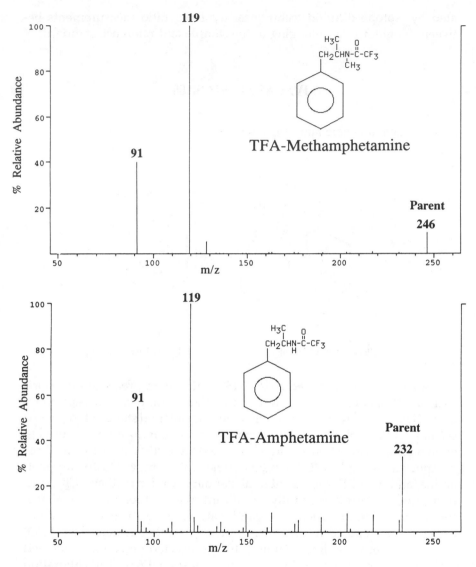

Figure 3.
Daughter ion spectrum of MPTFA-derivatized methamphetamine and amphetamine under CI(CH$_4$)/CAD(Ar). Parent ion is indicated.

phetamine and m/z 140/144 for amphetamine. Under such conditions the detection limit was found to be 0.5 ng/mg hair.

In this laboratory the analysis of amphetamine and methamphetamine has been accomplished via derivatization of the hair extract with *N*-methyl-bis(trifluoroacetamide) (MBTFA) followed by capillary GC/MS/MS using argon as a collision gas (Figure 3). For concurrent quantification, internal deuterated standards were employed: d_6-methamphetamine and d_5-amphetamine. In the case of amphetamine the parent ion selected for collision was the protonated molecular ion at m/z 232 produced by methane PICI. The daughter ions at m/z 91 and 119 were monitored for confirmation of presence. In the case of methamphetamine, the parent ion selected for collision was the protonated molecule ion at m/z 246 monitoring the daughter ions at m/z 91 and 119 for confirmation of presence. In this procedure the limits of quantification were as follows: 0.2 ng/10 mg hair for amphetamine and 0.5 ng/10 mg hair for methamphetamine. The percentage of metabolite amphetamine to the parent drug, methamphetamine, ranged from 3 to 20%.

The essential aspects of the extraction and derivatization procedures are as follows. The hair digest (0.5 ml), rendered alkaline (pH 10 to 11) with 1 ml of saturated sodium carbonate, is extracted with 4 ml of n-butyl chloride. The organic layer is extracted with 2 ml of 0.1 M sulfuric acid. The acid layer, made alkaline with 200 µl of 10 M NaOH, is re-extracted with 3 ml of n-butyl chloride. The organic layer is evaporated just to dryness in a Reacti-vial in the presence of 50 µl methanol containing 1% HCl as keeper. Derivatization with MBTFA (50 µl) is performed for 30 min at 70°C. The contents were evaporated at room temperature under nitrogen and reconstituted in 15 µl of toluene; 2 µl are injected on the column (Lauber injector).

B. Phencyclidine

PHENCYCLIDENE

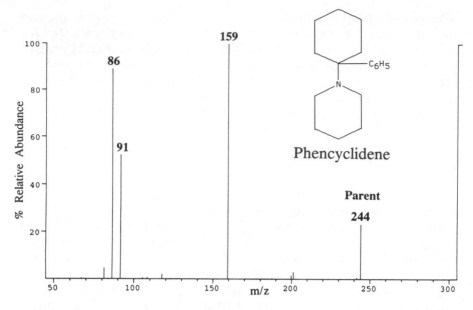

Figure 4.
Daughter ion spectrum of phencyclidine under CI(NH$_3$)/CAD(Ar). Parent ion is indicated.

Kidwell described[34,35] a pyrolysis analysis procedure for phencyclidine in hair samples. His approach avoided the extraction process by placing the sample into the direct insertion probe of the mass spectrometer for analysis. By employing isobutane CI, the protonated molecular ion, (M + H)$^+$, of any phencyclidine present would be favored, m/z 244. By selecting this ion for subsequent collision with argon, the daughter ions at m/z 159 and 86 were formed. While this procedure eliminated the extraction step and the chromatographic separation common to most analytical methods, the use of MS/MS still maintained the basic minimum elements of confirmation, namely the presence of three structurally related ions. The sensitivity of the method was determined to be 50 pg/mg.

In this laboratory the analysis of phencyclidine has been accomplished by capillary GC/MS/MS using ammonia for PICI and argon as a collision gas (Figure 4). For concurrent quantification the internal deuterated standard of d$_5$-phencyclidine was employed. The parent ion selected for collision was the protonated molecular ion at m/z 244 monitoring the daughter ions at m/z 86, 91, and 159 for confirmation of presence. In this procedure, the limit of quantification was 0.2 ng/100 mg hair.

The extraction procedure is as follows. Hair digest (0.5 ml) rendered alkaline (pH 10 to 11) with 1 ml of saturated sodium carbonate, is extracted with 4 ml of hexane. The hexane is back-extracted with 2 ml of 0.1 M sulfuric acid. The acid layer, made alkaline with 200 μl of 10 M NaOH (pH above 10), is re-extracted with 3 ml of hexane. After evaporating hexane under nitrogen at 40°C, the extract is reconstituted in 15 µl of toluene; 2 µl are injected on column (Lauber injector).

C. Morphine and Codeine

CODEINE   MORPHINE

With hair samples from heroin addicts, Pelli and Traldi[36] applied MS/ MS techniques via heated probe introduction. Using EI they demonstrated the detection and confirmation of morphine at levels as low as 10 pg/mg at a signal-to-noise level of 5:1. The selected parent ion, the molecular ion of morphine at m/z 285, when in collision with argon, produced four structural significant daughter ions at m/z 256, 242, 228, and 215. Without extraction and subsequent chromatographic separation, these authors had provided a protocol for confirmation of the presence of morphine in hair.

In a study on comparative determination of morphine in human hair using RIA and GC/MS, Sachs and Arnold[37] concluded that it was necessary to have confirmation by mass spectrometry. However, their criterion for confirmation and quantification was based on only one ion — m/z 246 — for morphine heptafluorobutyrate using the ion at m/z 479 from the internal standard levallorphan.

In this laboratory the analysis of morphine and codeine involves derivatization with N-methyl-N-trimethyl-silyltrifluoroacetamide (MSTFA) followed by capillary GC/MS/MS using ammonia as reagent gas and argon as a collision gas (Figure 5). For concurrent quantification, internal deuterated standards were employed: d_3-morphine and d_3-codeine. In the case of morphine, the parent ion selected for collision was the NH_3

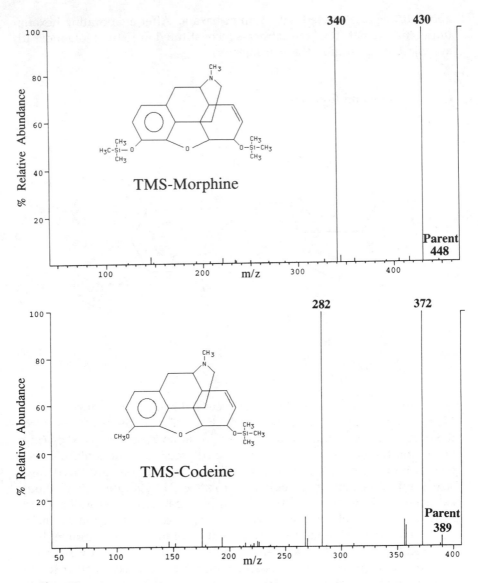

Figure 5.
Daughter ion spectrum of MSTFA-derivatized morphine and codeine under CI(NH₃)/
CAD(Ar). Parent ion is indicated.

molecular adduct at m/z 448 monitoring the daughter ions at m/z 340,
and 430 for confirmation of presence. In the case of codeine, the parent ion
selected for collision was the NH₃ molecular adduct at m/z 389 monitoring the daughter ions at m/z 282, and 372 for confirmation of presence.

The limits of quantification were 1 ng/10 mg hair for both morphine and codeine.

The extraction of analytes from 0.1 to 0.5 ml of hair digest is performed with Bond Elut Certify columns according to manufacturer's specifications. The column eluate (2 ml of freshly prepared methylene chloride to isopropyl alcohol (80:20) with 2% ammonium hydroxide) is evaporated under nitrogen, redissolved in 50 µl of methylene chloride, and transferred to an autosampler vial. After evaporation under nitrogen at 40°C, extracts are derivatized with 10 µl of MSTFA for 1 h at 70°C; 2 µl are injected on column (Lauber injector).

This laboratory does not currently measure the heroin metabolite monoacetylmorphine (MAM) because of the possibility of forming this by nonmetabolic hydrolysis reactions. The metabolite of choice for distinguishing between drug use and external contamination is morphine glucuronide. This metabolite, however, is not measured directly but by the increase in free morphine as a result of treating the hair digest with glusulase (New England Nuclear). The hydrolysis reaction is performed with 100 µl glusulase/1 ml of digest (10 mg hair per ml) for 3 h at 37°C and pH 5.2. The percentage increase of free morphine resulting from glusulase treatment ranges from 30 to 170%.

D. Cocaine, Benzoylecgonine, and Cocaethylene

Using a capillary column for chromatographic separation, Balabanova and Homoki[38] developed an assay for cocaine in hair samples based on GC/MS/EI. For confirmation purposes, these authors employed a full mass spectral scan for direct comparison with a reference spectrum. Under EI, the molecular ion for cocaine was present at m/z 303 together with five other prominent fragment ions, m/z 272, 198, 182, 105, and 82. The detection limit quoted for this approach was 100 fg. In this case history, the highest level of confirmation had been employed, that of full mass spectral scans. A similar approach was adopted and refined by Cone et al.[39] in testing human hair for unique cocaine metabolites of drug abusers. Moller et al.[40] used similar methods to identify and quantify cocaine and its metabolites in the hair of Bolivian coca chewers.

Martz[41] used a GC/MS/MS approach with deuterated analogs of cocaine (d_3) and benzoylecgonine (d_5) for stable isotope quantification. Methane CI was used to enhance the protonated molecular ions for these drugs for subsequent collision with argon, yielding daughter ions for confirmation. This approach was applied by Martz et al.[42] in the case of hair samples taken at the autopsy of a man who died from accidental cocaine poisoning. Fritch et al.[43] determined cocaine levels in hair of 132 individuals in a study to evaluate the results of RIA vs. GC/MS.

More recently, Curcuruto et al.[44] devised an ion trap mass spectrometry (ITMS) procedure whereby daughter spectra derived from the molecular ion for cocaine and morphine, m/z 303 and 285, respectively, were used for confirmation of presence. The daughter spectrum for cocaine contained the three ions for confirmation, m/z 198, 182, and 83. Analytical methods based on the ion trap detector have also recently been developed using full scan data for confirmation of presence and quantification.[45]

In this laboratory benzoylecgonine was analyzed by derivatization with MSTFA. Cocaine, cocaethylene, and trimethylsilyl benzoylecgonine were subjected to isobutane CI followed by CAD using argon as a collision gas (Figure 6). For concurrent quantification, internal deuterated standards were employed: d_3-cocaine, d_3-benzoylecgonine, and d_3-cocaethylene. For cocaine, the parent ion selected for collision was the protonated molecular ion at m/z 304 monitoring the daughter ions at m/z 82, and 182 for confirmation of presence. With derivatized benzoylecgonine, the parent ion selected for collision was the protonated molecular ion at m/z 362 monitoring the daughter ions at m/z 82, and 240 for confirmation of presence. For cocaethylene, the parent ion selected for collision was the protonated molecular ion at m/z 318 monitoring daughter ions at m/z 82, and 196 for confirmation of presence. In this procedure the limits of quantification were 2 ng/10 mg hair for cocaine, 0.5 ng/10 mg hair for benzoylecgonine, and 0.5 ng/10 mg hair for cocaethylene. The percentage of benzoylecgonine (BE/(BE + Cocaine) × 100) in typical hair samples is 16.5 ± 6.3 (1 SD), after corrections for *in vitro* hydrolysis.

Extraction and derivatization are performed as follows: hair digest (0.1 to 1.0 ml), adjusted with 0.1 M phosphate buffer to pH 6, is extracted on Analytichem Bond Elut Certify Columns according to the manufacturer's specifications. All three analytes are eluted with freshly prepared methylene chloride:isopropyl alcohol (80:20) with 2% ammonium hydroxide. Eluent

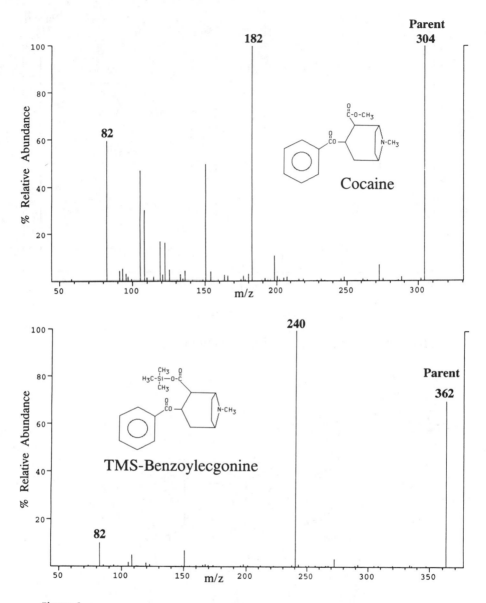

Figure 6.
Daughter ion spectrum of cocaine, cocaethylene, and MSTFA-derivatized
benzoylecgonine under $CI(C_4H_{10})/CAD(Ar)$. Parent ion is indicated.

is evaporated under nitrogen at 40°C and derivatized with 20 µl MSTFA
at 70°C for 1 h; 2 µl of derivatization mixture are injected on column
(Lauber injector).

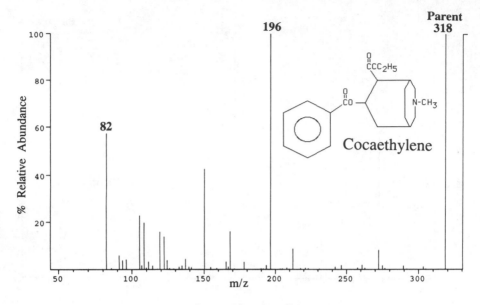

Figure 6. (continued)

E. 11-nor-9-Carboxy-9-Tetrahydrocannabinol

11-Nor-9-carboxy-Δ9-THC

In a recent study on analysis of THC in hair, Hayes et al.[46] employed a deuterated analog (d_3) of C-THC as an internal standard before derivatization with heptafluorobutyric anhydride (HFBA) and hexafluoroisopropanol (HFIP). Using a capillary column for chromatographic separation and NCI with ammonia as the reagent gas, the parent

HFBA/HFIP-11-nor-9-carboxy-9-tetrahydrocannabinol

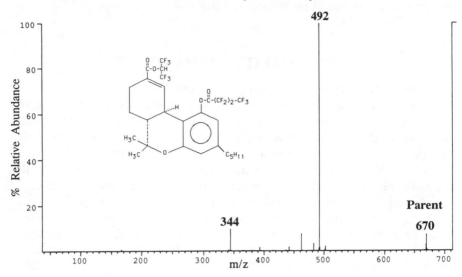

Figure 7.
Daughter ion spectrum of HFBA/HFIP-derivatized 11-nor-9-carboxy-9-tetrahydro-
cannabinol under NCI(NH$_3$)/CAD(Ar). Parent ion is indicated.

ion at m/z 670 was selected for CAD. Confirmation was then provided by
two daughter ions at m/z 344 and 492 obtained by collision with argon
(Figure 7). This method has been routinely used in our laboratory in several
thousand analyses. The results range from 0.3 pg to 10 pg/10 mg hair.

As much as half of the C-THC can be found in the melanin fraction of
the hair digest, even though the latter constitutes only about 5% of the hair
mass. Therefore, extraction procedures for all three possibilities have been
developed: intact hair, hair digest minus melanin, and melanin.

C-THC is extracted from the digest minus the melanin fraction as
follows. The digest (0.5 to 1.0 ml) is deproteinated by precipitation with 2
ml of acetonitrile. The supernatant is concentrated to 1.0 ml by evaporation under nitrogen at 40°C. The supernatant is made alkaline with 1 ml
0.2 N NaOH and extracted with hexane:ethyl acetate (9:1 v/v). The aqueous phase is acidified (1 ml 1 N HCl) and extracted with 2 ml hexane:ethyl
acetate (9:1). The organic phase is evaporated to dryness at 40°C with
nitrogen and derivatized at 70°C for 1 h with HFBA in the presence of
HFIP. It is important for the success of the derivatization to add HFIP
before the HFBA. Reagents are blown to dryness under nitrogen at room
temperature and reconstituted in 15 µl of toluene; 2 µl are injected on
column (Lauber injector).

Intact hair (5 to 10 mg) or the melanin fraction of the digest is extracted
with 200 µl of 10 N NaOH in 2 ml of deionized water in the presence of 200

µl of ethanol. Cleanup and derivatization procedures are identical to those described above except for deletion of the acetonitrile step.

V. FORENSIC STATUS OF HAIR TESTING

The forensic status of hair analysis can best be ascertained by a comparison of the probative features of hair and urine analysis. The latter, of course, enjoys the advantage of historic precedence but, as demonstrated here, not probative pre-eminence. Even the time element places the hair test in a stronger probative position relative to urine. Thus, the hair test can vindicate the results of urinalysis, but only additional hair testing with a newly collected sample can vindicate an earlier hair analysis result.

A simple example can make this clear. If a positive urinalysis result is challenged on the basis of specimen contamination or subversive activity such as spiked food or drink, a subsequently performed hair test can support or refute the proffered explanation for the positive urine result. However, if a positive hair test is challenged (e.g., mixed-up sample), only the analysis of a second, newly collected hair sample, but not a urine sample collected subsequent to the first hair specimen, can support or refute the first hair analysis result. Thus, the safety, or certainty, of urinalysis is greatly enhanced by hair analysis. The converse, however, does not apply.

In spite of its obvious supportive role for urinalysis, hair analysis continues to be challenged. Surprisingly, these challenges do not relate to the interpretation of a positive urine/negative hair result or to the more sophisticated mass spectrometry used in hair analysis, but rather to the clinical interpretation of what the critics call a "stand-alone" positive hair analysis result. In short, the critics maintain that at the present stage of development of the field, a positive hair analysis result on its own is not sufficient evidence of drug use. To distinguish drug use from external contamination, the critics maintain that hair analysis requires other competent evidence of drug use. At the same time, however, the critics agree that hair analysis meets all the requirements of forensic mass spectrometry.[47]

In essence, the "stand alone" issue raised against hair analysis employs the strategy of attacking the weaker elements of the forensic evidence, i.e., the interpretation or the weight of the evidence. This relatively weaker position of mass spectrometric data interpretation clearly is not unique to hair analysis. For one, such clinical interpretations rely on the laws of biochemistry and biology, and these, of course, provide us with less certain information than the laws of physics and chemistry forming the basis of the mass spectrometric criteria of identification and quantification. And this is true for hair as well as for urinalysis.

By attacking hair analysis in this manner, the critics have overlooked the fact that positive urinalysis results are by far more difficult to interpret than those of hair analysis. For instance, whether a positive urinalysis result is due to poppy seed ingestion, passive drug exposure, spiked food or drink, or willful drug use is, as shown above, a very difficult problem for urinalysis but not for hair analysis. Endogenous urine cutoff levels are an insecure guide in this respect. Consequently, it was mandated by NIDA that urinalysis not be a stand-alone test, but that positive results be supported by the clinical evaluation of a medical review officer. It is, therefore, faulty logic to claim forensic status for the probatively weaker urine test, while denying this for the probatively stronger hair test.

But even the characterization of hair analysis as a stand-alone test is incorrect. For if urine is not a stand-alone test, then hair analysis is even less so. For one, all hair analysis results, particularly the contested ones, are extensively reviewed by a toxicological/medical review officer who generally requests the collection and testing of a second ("safety net") hair sample. Moreover, this sample (as well as the original) is subjected to special testing and data evaluation. This includes: matching of first and second specimens by appearance and microscopic examination; evaluation of the hair specimen for evidence of hair treatment by methylene blue staining; and choosing an appropriate wash solution and appropriate segmentation of hair on the basis of methylene blue staining results. Additionally, positive mass spectrometric data is interpreted according to endogenous cutoff levels, the presence of metabolites and their cutoff values, wash kinetic criteria and their cutoff values, and reproducibility of results between the first and the second specimen. The latter frequently involves establishing the degree of agreement (between the first and second specimen) of the pattern of drug use as determined by segmental analysis. Such an extensive workup and evaluation by toxicological and medical review officers can hardly be characterized as a stand-alone test.

Although the probative weaknesses of urinalysis have long been recognized, they have not constituted grounds for excluding urinalysis from workplace testing. For one, it must be recognized in case of applicant testing that the employer receives the benefit of the doubt; for after all, the employer has the right to deny employment to individuals merely on the basis of such subjective criteria as unsuitable appearance or demeanor. Consequently, the employer has the right to adopt a policy of not only denying employment to individuals who use drugs but also to those who are in contact with drugs even though they may not be engaged in their use, e.g., a nondrug-using dealer who generates a positive urine result because of passive exposure to drugs. In essence this policy has also been adopted in the criminal justice system and in testing of military personnel during the development phase of urinalysis. Consequently, this policy should also be available to those wishing to engage in pre-employment testing by hair analysis.

When it comes to unannounced urine testing of employees in safety-sensitive positions, then the public receives the benefit of the doubt. However, safety-sensitive employees who produce a positive urine, but who deny drug use and do not exhibit any other evidence of drug use, should be given additional consideration not generally offered to applicants before being subjected to disciplinary action. For example, they may be given additional unannounced urine tests. The best approach, of course, would be for the medical review officer to request a hair test.

The probative power of hair analysis is now widely recognized by the relevant scientific community in over 130 independent scientific publications. This includes independent demonstrations of the effectiveness of wash procedures for removing external drug contaminants,[48] the absence of racial bias,[6,49] and the validation of hair analysis through self reports and concomitantly performed urinalysis.[7,8,50] The National Institute of Standards and Technology has organized three successful rounds of quality control surveys with 12 participating laboratories and now also offers quality control hair samples.

In 1990 the noted legal authority on scientific evidence, the Honorable Judge Jack B. Weinstein, set an important legal precedent in favor of hair analysis.[51] Currently at least 12 forensic laboratories, including those of the FBI, are using hair analysis for drugs of abuse in hundreds of forensic investigations.[26,31,36,40,42,43,52-57] We believe that these developments clearly indicate that hair testing has moved from its clinical origins to the forensic world, and this includes forensic quality workplace drug testing. Concerning future developments in the analysis of organic chemicals in hair, we believe that technical advances in instrumentation will play an important role, not only for such challenging substances as lysergic acid (LSD) or low concentration metabolites of drugs of abuse, but also for such relatively unexplored analytes as pharmacological agents and their metabolites as well as organic environmental toxins. In this respect, low-cost ion trap technology is likely to replace expensive triple quadrupole mass spectrometry when MS/MS becomes available in the next generation of ion trap instruments.

REFERENCES

1. Baumgartner, A. M., Jones, P. F. (Aerospace Corporation), Baumgartner, W. A., and Black, C. T. (Wadsworth V.A. Medical Center), *J. Nucl. Med.*, 20(7), 749, 1979.
2. Baumgartner, W. A., Hill, V. A., and Blahd, W. H., *J. Forensic Sci.* 34(6), 1433, 1989.

3. Higuchi, R., Von Broldingen, C., Sensabaugh, G. H., and Erlich, H. A., *Nature*, 318(6164), 543, 1988.
4. Cartmell, L. W., Aufderhide, A., and Weems, C., *J. Okla. State Med. Assoc.*, 84(1), 11, 1991.
5. Baumgartner, W. A. and Hill, V. A., Hair analysis for drugs of abuse: some forensic and policy issues, in *Proc. of Hair Analysis Conference*, National Institutes on Drug Abuse Monographs, in press.
6. Baumgartner, W. A. and Hill, V. A., Hair analysis for drugs of abuse: forensic issues, paper presented at *Proc. Int. Symp. Forensic Toxicology*, Federal Bureau of Investigation, Quantico, VA, June 15 to 19, 1992, in press.
7. Baer, J. D., Baumgartner, W. A., Hill, V. A., and Blahd, W. H., *Fed. Probation*, 55, 3, 1991.
8. Mieczkowski, T., Barzelay, D., Gropper, B., and Wish, E., *J. Psychoactive Drugs*, 23, 241, 1991.
9. Baumgartner, W. A., Testimony before the Subcommittee on Human Resources, Committee on Post Office and Civil Service, United States House of Representatives, May 20, 1987.
10. National Drug Control Strategy, Sept. 1989, *The White House National Priorities: A Research Agenda*, p. 83.
11. Mieczkowski, T., Landress, H. J., Newel, R., and Coletti, S. D., with comments by DeWitt, C. B., Director of National Institutes of Justice, *National Institute of Justice Research in Brief*, January 1993.
12. United States General Accounting Office, *Report on Drug Use: Strengths, Limitations, and Recommendations for Improvements*, June 1993, Report # GAO/PEND-93-18.
13. United Nations Interregional Crime and Justice Research Institute Workshop on Studying Epidemiology of Drug Use by Hair Analysis, in *Proceedings of the First International Meeting on Hair Analysis as a Diagnostic Tool for Drugs of Abuse Investigation*, Forensic Science International, in press.
14. Baumgartner, W. A. and Hill, V. A., Hair analysis for drugs of abuse, in *Recent Developments in Therapeutic Drug Monitoring and Clinical Toxicology*, Sunshine, I., Ed., Marcel Dekker, New York, 1992, 577.
15. Callahan, C. M., Grant, T. M., Phipps, P., Clark, G., Novack, A. H., Streissguth, A. P., and Raisys, V. A., *J. Pediatr.*, 120(5), 763, 1992.
16. Selavka, C. M., *J. Forensic Sci.*, 35, 685, 1991.
17. Baselt, R. C., Yoshikawa, D. M., and Chang, J. Y., *Clin. Chem.*, 37, 2160, 1991.
18. Harkey, M. R. and Henderson, G. L., Hair analysis for drugs of abuse. A critical review of the technology, *Report for the Department of Alcohol and Drug Programs*, Sacramento, CA, 1988.
19. Robbins, C. R., *Chemical and Physical Behavior of Human Hair*, 2nd ed., Springer-Verlag, New York, 1988, 59.
20. Baumgartner, W. A. and Hill, V. A., Sample preparation techniques, in *Proceedings of the 1st International Meeting on Hair Analysis as a Diagnostic Tool for Drugs of Abuse Investigation*, Forensic Science International, in press.
21. Baumgartner, W. A., Jones, P. F., Black, C. T., and Blahd, W. H., *J. Nucl. Med.*, 23(9), 790, 1982.
22. Valente, D., Cassini, M., Pigliaphchi, M., and Vansetti, G., *Clin. Chem.*, 27(11), 1952, 1981.
23. Baumgartner, A. M., Jones, P. F., and Black, C. T., *J. Forensic Sci.*, 23(9), 576, 1981.
24. Puschel, K., Thomasch, P., and Arnold, W., *Forensic Sci. Int.*, 21, 181, 1983.
25. Nagai, T., Kamiyama, S., and Nagai, T., *Z. Rechtsmed.*, 101, 151, 1988.
26. Kintz, P., Ludes, B., and Mangin, P., *J. Forensic Sci.*, 37, 328, 1992.
27. Cairns, T., Siegmund, E. G., and Stamp, J. J., *Mass Spectrom. Rev.*, 8, 93, 1989.

28. Stout, S. J., Steller, W. A., Manuel, A. J., Poeppel, A. R., and DaCunha, J., *J. Assoc. Off. Anal. Chem.*, 67, 142, 1984.
29. Cairns, T. and Siegmund, E. G., *Rapid Commun. Mass Spectrom.*, 1, 108, 1987.
30. Cairns, T., Siegmund, M. A., Luke, M. A., Doose, G. M., and Froberg, J. E., *Anal. Chem.*, 59, 2055, 1987.
31. Suzuki, O., Hattori, H., and Asano, M., *J. Forensic Sci.*, 29, 611, 1984.
32. Nakahara, K., Takahaski, K., Shimamine, M., and Takeda, Y., *J. Forensic Sci.*, 46, 70, 1990.
33. Nakahara, K., Takahaski, K., Takeda, Y., Kyohei, K., and Tokui, T., *J. Forensic Sci.*, 46, 243, 1990.
34. Kidwell, D., Analysis of drugs of abuse in hair by tandem mass spectrometry, *36th Conf. Am. Soc. Mass Spectrom.*, San Francisco, 1988.
35. Kidwell, D. A., *J. Forensic Sci.*, 38, 272, 1993.
36. Pelli, B. and Traldi, P., *Biomed. Environ. Mass Spectrom.*, 14, 63, 1987.
37. Sachs, H. and Arnold, W., *J. Clin. Chem. Clin. Biochem.*, 27, 873, 1989.
38. Balabanova, S. and Homoki, J., *Z. Rechtsmed.*, 98, 235, 1987.
39. Cone, E. J., Yousefnejad, D., Darwin, W. D., and Maguire, T., *J. Anal. Toxicol.*, 15, 250, 1991.
40. Moller, M. R., Fey, P., and Rimbach, S., *J. Anal. Toxicol.*, 16, 291, 1992.
41. Martz, R. M., *Crime Lab. Dig.*, 15, 67, 1988.
42. Martz, R., Donnelly, B., Fetterolf, D., Lasswell, L., Hime, G. W., and Hearn, W. L., *J. Anal. Toxicol.*, 15, 279, 1991.
43. Fritch, D., Groce, Y., and Rieders, F., *J. Anal. Toxicol.*, 16, 112, 1992.
44. Curcuruto, O., Guidulgli, F., Traldi, P., Sturaro, A., Tagliaro, F., and Marigo, M., *Rapid Commun. Mass Spectrom.*, 6, 434, 1992.
45. Harkey, M. R., Henderson, G. I., and Zhou, C., *J. Anal. Toxicol.*, 15, 260, 1991.
46. Hayes, G., Scholtz, H., Donahue, T., and Baumgartner, W., *39th Conf. Am. Soc. Mass Spectrom.*, Nashville, 1991.
47. Hearn, W. L., Chairman, Advisory Committee on Hair Analysis for Drugs of Abuse, Report presented to the Annual Meeting of the Society of Forensic Toxicologists, October 13 to 17, 1992, Cromwell, CT.
48. Koren, G., Klein, J., Forman, R., and Graham, K., *J. Clin. Pharmacol.*, 32, 671, 1992.
49. Mieczkowski, T. and Newel, R., *Forensic Sci. Int.*, in press.
50. Magura, S., Freeman, R. C., Siddiqi, Q., and Lipton, D. S., *Int. J. Addictions*, 27, 51, 1992.
51. U.S. *vs.* Antony Medina, U.S. District Court, Eastern District of New York, Docket #87-CR-824-3, October 22, 1990.
52. Smith, F. P. and Liu, R. H., *J. Forensic Sci.*, 31(4), 1269, 1986.
53. Arnold, W., *Forensic Sci. Int.*, 46, 17, 1990.
54. Klug, E., *Z. Rechtsmed.*, 84, 189, 1980.
55. Arnold, W. and Sachs, H., *Bull. Int. Assoc. Forensic Toxicol.*, 22(4), 9, 1992.
56. Staub, C., *Forensic Sci. Int.*, in press.
57. Ishiyama, I., Nagai, T., and Toshida, S., *J. Forensic Sci.*, 28(2), 380, 1983.

Chapter **3**

MASS SPECTROMETRY IN SPORTS TESTING

Jongsei Park, Songja Park, Dongseok Lho, and Bongchul Chung

CONTENTS

0-8493-8252-1/95/$0.00+$.50
© 1995 by CRC Press Inc.

I. INTRODUCTION

The history of mankind is the history of man's endeavor to improve his performance and thereby his chance of winning in combat and sports. Thus, there is a long story of administrations that would today be called doping and of endeavors to eliminate such methods.[1]

Even the athletes in the ancient Olympic Games at the end of the 3rd century B.C. are said to have tried to improve their achievements by any means possible. Detailed reports have come from Latin and South American areas, where different stimulants from harmless maté, tea, and coffee up to strychnine and cocaine were used to increase performance and to deaden a feeling of hunger during long marches. Europe only learned the use of caffeine-containing drugs towards the end of the 16th century. However, only since the second half of the 19th century are there verified examples of unquestionable doping in sports.

Many different sports were known to be touched by doping scandals. It included swimmers, cycling racers, sprinters, soccer players, and boxers — in many cases with trainers' help.

Human medical science discovered an interest in medicines and substances which could improve performance and tested these in a comprehensive series of experiments. Pharmacological effects of benzedrine, methedrine, and amphetamines were tested for the purpose of doping use.

It required several fatalities to make public opinion aware of the urgency of the doping agents abuse in sports. The best known is the case of the Danish cyclist Knut Enemark Jensen who collapsed at the 1960 Olympic Games in Rome after a fatal dose of amphetamines and nicotinic acid compounds which were given to him by his trainer. In 1957, the American Medical Association reported that amphetamine "pep pills", used originally by students before examinations, had found wide usage in sports, too.

The growing commercialization of sport explains the increasingly close connection between doping and professionalism. Thus, doping with amphetamines and anabolic steroids has also become more prevalent in football.

A particular official impetus for the war on doping came from the Council of Europe, which convened an expert commission in Strasbourg in January 1963. Since the first attempt to implement antidoping measures had shown that the greatest problem for efficient antidoping action would be a clear definition of the term "doping", such a definition was developed at the outset and reads as follows: "Doping is defined as the administering or use of substances in any form alien to the body or of physiological substances in abnormal amounts and with abnormal methods by healthy persons with the exclusive aim of attaining an artificial and unfair increase of performance in competition. Furthermore various psychological measures to increase performance in sports must be regarded as doping."

Very significant in sports was the establishment in 1967 of the IOC Medical Commission, which conducted its first routine examinations at the 1968 Olympics in Grenoble and continued its work during the 1968 Games of the XIXth Olympiad in Mexico. Because of greatly differing interpretations of the List of Doping Substances, the standards applied to the individual athletes vary considerably and do not always accord with the rules of the Olympic Games. Nevertheless, in collaboration with the international sports federations, the IOC conducted nearly 16,000 doping tests from 1968 to 1992 in the Olympic Games alone and disqualified those violating the regulations. The first list of doping classes was established by the IOC-Medical Commission in 1968 and it has been changed from time to time to reflect changes of doping abuse in sports.

The first list subdivided doping agents into four groups according to their pharmacological effects, i.e., psychomotorically stimulating substances, sympathomimetic substances, substances stimulating the CNS, and narcotics and strong analgesic agents. In 1974, anabolic hormones were added as a fifth group.

A list of doping classes and methods by the IOC as of 1993 is presented in Table 1. This is not a final list, but should include all pharmacologically similar substances. This is a necessity, because it is possible, by means of small, easily executable chemical variations, to produce numerous substances with the same constant pharmacological effect.

This list has, in the meantime, been accepted without any limitation by all the international sports federations. This led to the unification of doping regulations, which was needed as a basis for joint international action. Because of this, there has been no room for excuses for the use of doping agents for a number of years already. Therefore, those interested in doping opportunities are concentrating their efforts more and more on the discovery surety, and on the camouflage, with the aid of other substances, to make identification impossible. But there is no chance for such intentions today.

TABLE 1.

List of Doping Classes and Methods by IOC (March 1993)

List of Doping Classes and Methods

Doping classes
   Stimulants
   Narcotics
   Anabolic Agents
   Diuretics
   Peptide hormones and analogs
Doping methods
   Blood doping
   Pharmacological, chemical, and physical manipulation
Classes of drugs subject to certain restrictions
   Alcohol
   Marijuana
   Local anesthetics
   Corticosteroids
   Beta-blockers

Note: The doping definition of the IOC is based on the banning of pharmacological classes of agents.

The hottest problem for many years was the use of anabolic hormones. The progress in analytical methods and the possibility of the identification of anabolic substances was a decisive step in the history of the fight against doping.

Satisfactory progress has been made in the fight against doping abuse during the last few years. Effective publicity is still the most important demand in the struggle against doping as a form of deceit and self-deception.

II. ANALYSIS OF DOPING AGENTS

A. Overview of Analysis

Testing for doping substances at the Olympic Games started officially during the 1968 Grenoble (France) Winter Olympics. Due to the analytical techniques available at that time, only a few stimulants and narcotics were checked, these being the most widely abused substances among athletes at the time.[2] At the 1976 Montreal Olympic Games, anabolic steroids were added as screened substances. About 15% of the specimens were tested for anabolic steroids at that time. During the 1980 Moscow Olympic Games, half of the specimens were tested for anabolic steroids. Analytical techniques for the detection of anabolic steroids by gas chromatography/mass spectrometry (GC/MS) had advanced enough since that time so that all specimens were tested for anabolic steroids by GC/MS since the 1984 Los Angeles Olympic Games.[3]

1. Screening Procedures

The IOC-MC classified banned substances according to their pharmacological properties (refer to banned drugs lists). Many laboratories set up doping analysis methods to cover as many drugs as possible without sacrificing sensitivity and specificity within one procedure. Its screening method consisted of five different procedures; GC, GC/MSD, LC, and TDx were used, and every presumptive positive case was confirmed by GC/MS. Detailed analytical procedures have been reported.[4] The screening procedures are briefly described as follows:

- *Procedure I: volatile doping agents.* Alkaline extraction of urine by ether, with the ethereal layer injected onto a gas chromatograph equipped with a nitrogen specific detector and a 5% phenylmethylsilicone capillary column (17 m × 0.2 mm i.d., film thickness 0.33 μm) was used to screen for stimulants.

- *Procedure II: phenylalkylamines, narcotic analgesics, and beta-blockers.* The procedure involves acid hydrolysis, extraction at pH 9.6 with borate buffer, selective derivatization, and injection into a GC/MSD using a cross-linked methylsilicone capillary column (17 m × 0.2 mm i.d., film thickness 0.33 μm).

- *Procedure III: caffeine, pemoline, diuretics, and corticosteroids.* The procedure involves liquid-liquid extraction with ether, followed by evaporation, dissolving the residue in methanol, and injection into an HPLC using an ODS column with gradient elution and a diode-array UV detector.

- *Procedure IV: anabolic steroids.* The procedure involves extraction of urine with XAD-2, elution with methanol, separation of nonconjugated steroids from conjugated steroids between buffer and ether, derivatization of nonconjugated steroids without further treatment, and derivatization of conjugated steroids after enzymatic hydrolysis.

- *Procedure for amphetamines, opiates, cocaine metabolites, cannabinoids, barbiturates, and benzodiazepines.* To supplement GC and GC/MS data for doping analysis, a fluorescence polarization immunoassay (FPIA) was used.

The FPIA method is a homogeneous competitive immunoassay system utilizing fluorescence polarization immunoassay principles. The system can perform assays of therapeutic drugs, hormones, clinical chemistries, proteins, and abused drugs and their metabolites in serum or urine.

Methods to analyze amphetamines, opiates, cocaine metabolites, phencyclidine, cannabinoids, barbiturates, and benzodiazepines were available.

TABLE 2.

Doping Class of Stimulants

Amfepramone	Amfetaminil	Amineptine
Amiphenazole	Amphetamine	Benzophetamine
Caffeine[a]	Cathine	Chlorphentermine
Clobenzorex	Clorprenaline	Cocaine
Cropopamide (component of "micoren")		
Crothetamide (component of "micoren")		Dimethamfetamine
Ephedrine	Etafedrine	Ethamivan
Etilamfetamine	Fencamfamine	Fenetylline
Fenproporex	Furfenorex	Mefenorex
Mesocarb	Methamphetamine	Methoxyphenamine
Methylephedrine	Methylphenidate	Morazone
Nikethamide	Pemoline	Pentetrazol
Phendimetrazine	Phenmetrazine	Phentermine
Phenylpropanolamine	Pipradol	Prolintane
Propylhexedrine	Pyrovalerone	Strychnine
and related compounds		

[a] For caffeine the definition of a positive depends upon whether the concentration in urine exceeds 12 µg/ml.

B. Analysis of Stimulants and Narcotics

Stimulants (Table 2) comprise various types of drugs which increase alertness, reduce fatigue, and may increase competitiveness and hostility. Their use can also produce loss of judgment, which may lead to accidents to others in some sports. Amphetamine and related compounds have the most notorious reputation of producing problems in sport. Some deaths of sportsmen have resulted even when normal doses have been used under conditions of maximum physical activity. There is no medical justification for the use of "amphetamines" in sports.

1. Analysis of Stimulants

a. Analysis

A typical example of analysis of stimulants in urine is described as follows:[5,6]

- *Instrumentation.* A gas chromatograph equipped with a nitrogen-phosphorous detector (NPD) was used to screen stimulants in urine. Separation was achieved with a fused-silica capillary column with cross-linked 5% phenylmethylsilicone (SE-54), 17 m length, 0.2 mm i.d., 0.33-µm film thickness. The chromatographic conditions were as follows: detector, NPD at 300°C; injector temperature, 280°C; initial temperature,

100°C; ramp, 20°C/min; final temperature, 300°C for 4 min; carrier gas, helium at a flow of 1.5 ml/min: hydrogen flow, 3 ml/min; air flow, 70 ml/min; split ratio 1:10. For strychnine, anileridine, and doxapram, the conditions were modified as follows: initial temperature, 280°C; ramp, 20°C/min; final temperature, 330°C for 2 min. The mass spectrometric confirmation of a positive urine specimen was made with a gas chromatograph interfaced with a mass selective detector. A SE-54 column as used for the chromatographic separation. The chromatographic parameters of GC/MSD were as follows: detector, mass selective detector in scan mode; ionization, electron impact mode; ionization potential, 70 eV; injector temperature, 280°C; interface temperature, 300°C; initial temperature, 100°C; ramp, 20°C/min; final temperature, 300°C for 3 min; carrier gas, helium at a flow rate of 0.9 ml/min; split ratio, 1:10.

- *Extraction.* A 5-ml urine sample was transferred to a 15-ml glass-stoppered centrifuge tube. To this were successively added 0.5 ml of 5 N potassium hydroxide; 25 µl of methanolic solution of N,N-Diisopropyl-1-amino-n-dodecane as an internal standard, 1000 ppm; 2 ml of distilled diethylether; and 3 g of anhydrous sodium sulfate. The mixture was shaken mechanically for 20 min and then centrifuged at 2500 rpm. The organic layer was transferred with a glass syringe into a 2-ml chromatographic vial and submitted to gas chromatographic analysis.

- *Selective derivatization.* The etheral layer obtained as previously mentioned was evaporated to dryness in a vacuum rotary evaporator. The residue was dissolved with 50 µl of a mixture of acetonitrile:trifluoroacetic acid (60:40, v/v) which contained 200 ppm of methyl orange. To the solution, 60 to 70 µl of N-methyl-n-trimethylsilyltrifluoroacetamide (MSTFA) was added until the color changed from red to yellow. The mixture was heated to 80°C for 5 min with a heating block. To the solution, 10 µl of N-methyl-bis-trifluoroacetamide (MBTFA) was added and it was heated again to 80°C for 10 min. This solution was injected into the GC/MSD.

b. Results and Discussion

Stimulants and narcotic analgesics banned according to International Olympic Committee regulations are nitrogen-containing compounds. Most of these drugs are excreted in free form in urine at high concentrations and as their metabolites at low, often undetectable concentrations.[7] These drugs may be directly extracted from urine under alkaline conditions without hydrolysis. Most of stimulants were extracted with 70 to 90% recovery, but ethamivan was extracted with a poor yield (18%) because it is a phenolic compound containing an electron-attracting carbonyl group. The inductive effect of the carbonyl group on the aromatic ring increases the acidity of a phenolic hydrogen. Ethamivan should be extracted under mild basic conditions. The recovery data of caffeine, heptaminol, and norpseudoephedrine showed low yields of 30 to 50%. The poor reproducibility of caffeine (8.8% RSD) is due to the strong alkaline extraction. The quantitative analysis of caffeine is therefore performed using the solid-phase extraction or liquid-liquid extraction under neutral conditions.[8]

A GC chromatogram of a mixture of 40 drugs indicates good separation as shown in Figure 1. The GC analyses of samples serve only to indicate the possible presence of a drug, but do not provide definite proof. For this reason, mass spectrometric confirmation is required for doping control tests in sports. All confirmations by GC/MS used corresponding reference compounds. Table 3 shows the characteristic ions and relative retention times of the derivatives of stimulants and narcotics.[6]

2. Analysis of Narcotics

a. Analysis

A typical example of analysis of narcotics in urine is described as follows:[6,9,10]

- *Instrumentation.* A fused-silica capillary column SE-30 was used for gas chromatographic separation interfaced to a mass selective detector. The detector was operated in the scan mode. The electron impact ionization potential was 70 eV. The injector temperature was 280°C and the interface temperature was 300°C. The temperature program was as follows: an initial temperature of 160°C was increased by 20°C/min to a final temperature of 300°C and held for 3 min. The carrier gas was helium at a flow rate of 0.9 ml/min and the split ratio was 1:10.

- *Sample preparation.* A 5-ml urine sample was transferred into a 15-ml glass-stoppered centrifuge tube. To this were added 1 ml of 6 N hydrochloric acid and 100 mg of cysteine. The solution was heated to 105°C for 3 min on a heating block and then cooled to room temperature. To the solution, 5 ml of diethylether was added, and the mixture was mechanically shaken for 10 min and centrifuged at 2500 rpm for 5 min. The organic layer was discarded. To the aqueous fraction, 20 µl of a 0.5 mg/ml ψ-ephedrine solution was added as an internal standard for screening of phenolalkylamines and narcotic analgesics. The pH of the solution was adjusted to 9.6 with borate buffer. To the solutions, 5 ml of diethylether and 3 g of anhydrous sodium sulfate were successively added. The mixture was mechanically shaken for 20 min and centrifuged at 2500 rpm for 5 min. The ethereal fraction was transferred into a 15-ml glass-stoppered centrifuge tube and evaporated to dryness. The residue was dissolved with 50 µl of a mixture of acetonitrile:trifluoroacetic acid (TFA) (60:40, v/v) containing 200 ppm of methylorange. MSTFA was added dropwise to the solution until the color changed from red to yellow. The mixture was heated to 80°C for 5 min. To the solution, 10

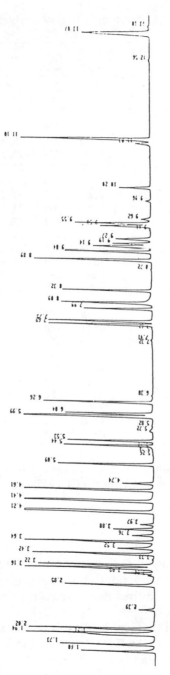

Figure 1.
Gas chromatography of a mixture of stimulants and narcotics. 1.60, heptaminol; 1.73, amphetamine; 1.91, phentermine; 1.94, propylhexedrine; 2.02, methylamphetamine; 2.39, dimethylamphetamine; 2.85, norpseudoephedrine; 3.08, chlorphentermine; 3.16, ephedrine; 3.22, methoxyphenamine; 3.42, methylephedrine; 3.52, p-hydroxyamphetamine; 3.64, phenmetrazine; 3.76, phendimetrazine; 3.88, pholedrine; 3.97, diethylpropion; 4.21, nikethamide; 4.41, cardiazole; 4.61, fenproporex; 4.74, metaraminol; 5.09, fencamfamine; 5.44, methylphenidate; 5.53, pethidine, I.S.; 5.99, caffeine; 6.04, benzphetamine; 6.26, ethamivan; 7.69, methadone; 7.75, pipradol; 7.99, cocaine; 8.09, levorphanol; 8.32, pentazocine; 8.89, codeine; 9.04, ethylmorphine; 9.19, hydrocodone; 9.55, oxycodone; 10.20, phenazocine; 11.03, anileridine; 11.10, doxapram; 13.07, strychnine.

TABLE 3.

Relative Retention Times and Characteristic Ions of Derivatives of Stimulants and Narcotics

No.	Derivative	RRT	Base peak	Characteristic ions
1	Amphetamine-NTFA	0.495	140	118, 91, 69
2	Anileridine-NTFA	1.803	246	247, 375, 42
3	Caffeine	1.000	194	109, 55, 67
4	Chlorphentermine-NTFA	0.693	154	59, 114, 89
5	Codeine-OTMS	1.460	371	73, 372, 178
6	Dihydrocodeine-OTMS	1.426	373	73, 146, 42
7	Ephedrine-NTFA-OTMS	0.733	179	73, 77, 45
8	Etafedrine-OTMS	0.735	86	58, 72, 116
9	Ethamivan-OTMS	1.100	223	73, 295, 224
10	Ethylmorphine-OTMS	1.490	385	73, 192, 146
11	Fencamfamine-NTFA	1.023	142	170, 91, 115
12	Fenproporex-NTFA	0.884	193	56, 140, 118
13	Heptaminol-NTFA-OTMS	0.582	131	73, 75, 69
14	Levorphanol-OTMS	1.288	59	73, 329, 328
15	Metaraminol-NTFA-(OTMS)$_2$	0.871	267	73, 45, 77
16	Methoxyphenamine-NTFA	0.761	153	148, 109, 42
17	Methamphetamine-NTFA	0.601	154	109, 42, 91
18	Methylephedrine-OTMS	0.719	72	162, 102, 191
19	Methylphenidate-NTFA	1.051	180	90, 150, 67
20	Norpseudoephedrine-NTFA-OTMS	0.648	179	73, 45, 77
21	Oxycodone-OTMS	1.553	387	388, 73, 372
22	Pentazocine-OTMS	1.327	289	342, 357, 274
23	Phenazocine-OTMS	1.597	302	303, 73, 58
24	Phenmetrazine-NTFA	0.774	70	167, 55, 98
25	Phentermine-NTFA	0.509	154	59, 91, 132
26	Pipradol-OTMS	1.243	84	73, 56, 165
27	Propylhexedrine-NTFA	0.570	154	110, 42, 182

µl of MBTFA was added and the solution was heated again to 80°C for 10 min. This solution was injected onto the GC/MSD.

b. Results and Discussion

i. Narcotic Analgesics

The majority of narcotic analgesics (Table 4) have a rigid molecular structure and therefore possess very poor chromatographic properties. Their chromatographic analysis requires derivatization prior to GC separation and detection.

The phenolic, allylic, and aliphatic hydroxy groups in narcotic analgesics were easily derivatized with MSTFA to form the corresponding trimethylsilyl ethers. No side products were observed. In mass spectra of these derivatives, the individual base peak was found to be the corresponding molecular ion (Table 5).

TABLE 4.

Doping Class of Narcotic Analgesics

Alphaprodine	Anileridine	Buprenorphine
Codeine	Dextromoramide	Dextropropoxyphene
Diamorphine (heroin)	Dihydrocodeine	Dipipanone
Ethoheptazine	Ethylmorphine	Levorphanol
Methadone	Morphine	Nalbuphine
Pentazocine	Pethidine	Phenazocine
Trimeperidine	and related compounds	

On trifluoroacylation of narcotic analgesics with MBTFA, the phenolic hydroxy group was converted to the OTFA derivative, but the allylic and aliphatic hydroxy groups were not. With morphine, trifluoroacylation gave di- and mono-OTFA derivatives, which were also observed in derivatization with trifluoroacetic anhydride. The aliphatic hydroxy group of oxycodone and oxymorphone did not react with MBTFA, but it did with MSTFA to form the corresponding TMS-ether. The trimethylsilylation technique was therefore used for screening of narcotic analgesics by GC/MS. Characteristic ions of the trifluoroacylated products of narcotic analgesics are listed in Table 6.

ii. Phenolalkylamines

n-Alkyl primary amines (tyramine, octopamine, and norfenefrine) and α-substituted primary amines (metaraminol, *p*-hydroxyamphetamine, and *p*-hydroxynorephedrine) were derivatized with MSTFA to form *N*-mono-TMS and *N*-di-TMS derivatives. The major products were a *N*-di-TMS derivative for primary amines and a *N*-mono-TMS derivative for α-substituted primary amines.

Trimethylsilylation of *N*-methyl secondary amines (pholedrine and phenylefrine) gave *N*-mono-TMS derivatives (major) and underivatized amines (minor). The *N*-ethyl(etilefrin) or *N*-butyl(bamethan) secondary amine was not trimethylsilylated, because a bulky group in an amino function hinders the introduction of a TMS group to the nitrogen atom. Fragmentation of phenolalkylamine-NTMS-OTMS derivatives takes place alpha to the nitrogen atom and yields an iminium cation from the loss of a benzyl group. The iminium ion is resonance stabilized and was found to be the base peak in mass spectra of the derivatives. Trimethylsilylation of phenolalkylamines did not produce a single product, and the mass spectra of the resulting derivatives indicated that the individual base peak depended upon the degree of derivatization of the amine function (Table 5). Therefore, this derivatization technique was not suitable for screening all phenolalkylamines.

Trifluoroacylation of phenolalkylamines with *N*-methyl-bis-trifluoroacetamide (MBTFA) resulted in the formation of the correspond-

TABLE 5.

Characteristic Ions of the Trimethylsilylated Phenolalkylamines and Narcotic Analgesics

Drug	Mol wt	Base peak	Characteristic ions (m/z)
P-OH-Amphetamine-NTMS-OTMS	295	116	161, 251, 280
Tyramine-N(TMS)$_2$-OTMS	353	174	86, 251, 338
Metaraminol-NTMS-(OTMS)$_2$	383	116	147, 267, 368
Norfenefrine-N(TMS)$_2$-(OTMS)$_2$	441	174	147, 352, 426
Pholedrine-NTMS-OTMS	309	58	115, 209, 294
p-OH-Norephedrine-NTMS-(OTMS)$_2$	311	267	81, 193, 296
Octopamine-N(TMS)$_2$-(OTMS)$_2$	441	174	100, 267, 426
Phenylephrine-NTMS-(OTMS)$_2$	398	116	147, 267, 373
Etilefrin-(OTMS)$_2$	325	58	267, 147, 310
Bamethan-(OTMS)$_2$	353	267	174, 207, 353
Phthidine	247	72	91, 117, 165
Levorphanol-OTMS	329	59	329, 272, 150
Methadone	310	72	296, 85, 200
Pentazocine-OTMS	357	289	245, 342, 357
Dihydrocodeine-OTMS	373	373	146, 236, 315
Codeine-OTMS	371	371	178, 234, 313
Hydrocodone	299	299	242, 214, 185
Ethylmorphinbe-OTMS	385	385	192, 146, 287
Morphine-(OTMS)$_2$	429	429	236, 146, 287
Hydromorphone-OTMS	357	357	300, 342, 243
Oxycodone-OTMS	387	387	229, 273, 330

ing NTFA-OTFA derivatives. MBTFA reacts primarily with the phenolic hydroxy and the amino groups, and then the unreacted MBTFA can react with the aliphatic hydroxy group. The difference between nucleophilicities of the phenolic and the aliphatic hydroxy groups towards MBTFA leads to the formation of mono- and di-OTFA derivatives. α-Fission of the N,O-trifluoroacylated phenolalkylamines also produce the resonance-stabilized iminium ion as the base peak at m/z 140 or 154 (Table 6). This derivatization did not produce a single product and was not suited for screening of phenolalkylamines.

N-Trifluoroacyl-O-trimethylsilyl (NTFA-OTMS) derivatives of phenolalkylamines were easily prepared by trimethylsilylation (OTMS) with MSTFA followed by selective N-trifluoroacylation.

The phenolalkylamine-NTFA-OTMS derivatives were relatively stable compounds at elevated temperature and against oxidizing agents. The selective derivatization produced a single product. In the mass spectra of all phenolalkylamine-NTFA-OTMS derivatives, the base peak at m/z 179 or 267 arose from the benzylic fragment (Table 7), thus yielding information on the hydroxylic substitution pattern of the aromatic ring and/or

TABLE 6.

Characteristic Ions of the Trifluoroacylated Phenolalkylamines
and Narcotic Analgesics

Drug	Mol wt	Base peak	Characteristic ions (m/z)
P-OH-Amphetamine-NTFA-OTFA	343	140	230, 69, 343
Tyramine-NTFA-OTFA	329	216	69, 126, 329
Metaraminol-NTFA-(OTFA)$_2$	455	140	69, 246, 359
Norfenefrine-NTFA-(OTFA)$_2$	441	232	69, 121, 345
Pholedrine-NTFA-OTFA	357	154	230, 69, 342
p-OH-Norephedrine-NTFA-(OTFA)$_2$	465	140	69, 219, 302
Etilefrin-NTFA-(OTFA)$_2$	469	154	69, 81, 356
Synephrine-NTFA-(OTFA)$_2$	455	140	69, 328, 342
Levorphanol-OTFA	353	353	285, 150, 115
Pentazocine-OTFA	381	314	298, 366, 381
Dihydrocodeine-OTFA	397	397	284, 185, 340
Codeine-OTFA	395	282	395, 266, 338
Morphine-(OTFA)$_2$	477	364	477, 380, 307
Hydromorphone-OTFA	381	381	325, 96, 296
Oxycodone-OH	315	315	230, 258, 201
Phenazocine-OTFA	409	269	296, 192, 409
Ethylmorphine-OTFA	409	409	296, 266, 352

benzylic carbon atom. The m/z 179 and 267 ions were used for screening of all phenolalkylamines by GC/MS. The total ion chromatogram of a mixture of 12 phenolalkylamines that were selectively derivatized showed a good separation. No by-products were observed in the chromatogram. The selective derivatization technique also provided a good separation of very closely related phenolalkylamines which had been difficult to resolve in the underivatized state.

iii. Simultaneous Determination of Phenylalkylamine
      and Narcotic Analgesics

Phenolalkylamines and narcotic analgesics can be simultaneously screened by GC/MSD using the multiple-ion monitoring mode; the m/z 179 and 267 ions were monitored. The retention time of a peak with ions m/z 179 and 267 was compared with that of a standard compound, and then the suspicious peak was identified by comparing its full spectrum with that of the corresponding standard.

For screening of narcotic analgesics, the molecular ion and two other characteristic ions were monitored. The retention times and abundances of peaks at the three monitored ions were compared with those of an authentic standard. The monitored ions for screening phenolalkylamines and narcotic analgesics are listed in Table 8.

TABLE 7.

Characteristic Ions of the NTFA-OTMS Derivatives of
Phenolalkylamines

Compound	Mol wt	Base	Characteristic ion (m/z)
p-OH-Amphetamine-NTFA-OTMS	319	179	206, 77, 319
Tyramine-NTFA-OTMS	305	179	193, 77, 305
Metaraminol-NTFA-(OTMS)$_2$	407	267	77, 197, 392
Norfenenfrine-NTFA-(OTMS)$_2$	393	267	77, 179, 378
Pholedrine-NTFA-OTMS	333	179	206, 154, 333
p-OH-Norephedrine-NTFA-(OTMS)$_2$	417	267	193, 236, 294
Octopamine-NTFA-(OTMS)$_2$	393	267	193, 305, 392
Phenylefrine-NTFA-(OTMS)$_2$	407	267	179, 302, 392
Etilefrin-NTFA-(OTMS)$_2$	421	267	177, 193, 406
Synefrine-NTFA-(OTMS)$_2$	407	267	193, 317, 392
p-OH-Ephedrine-NTFA-(OTMS)$_2$	421	267	193, 251, 407
Ethamivan-OTMS	294	223	294, 193, 264
Bamethan-NTFA-(OTMS)$_2$	449	267	179, 359, 434

C. Analysis of Anabolic Agents

In 1974, steroids were added to the list of doping agents banned by the International Olympic Committee on the grounds that drug abuse is against the Olympic spirit based on fair play and on the fact that steroid abuse will harm the well-being of athletes. Nevertheless, it is suspected that the use of steroids has been steadily increasing among the players, who believe that steroid use improves their physical strength remarkably over a short period of time. These athletes are willing to barter their health for fame and monetary gain.

The anabolic androgenic steroid (AAS, Table 9) class includes testosterone and substances that are related in structure and activity to it. They have been misused by the sports world both to increase muscle strength and bulk, and to promote aggressiveness. The use of AAS is associated with adverse effects on the liver, skin, cardiovascular, and endocrine systems. They can promote the growth of tumors and induce psychiatric syndromes. In males, AAS decreases the size of the testes and diminishes sperm production. Females experience masculinization, loss of breast tissue, and diminished menstruation. The use of AAS by teenagers can stunt growth.

1. Analysis of Anabolic Agents

Some steroids can be analyzed by high-performance liquid chromatography (HPLC).[11] It has been possible to detect 17α-ethyl and 19-nortestosterone steroids by radioimmunoassay despite the method's low specificity.[12] However, both these methods cannot provide the data neces-

TABLE 8.

Selected Ions for Screening Phenolalkylamines and
Narcotic Analgesics

Compound	Extracted ion (m/z)
Phenolalkylamines-NTFA-OTMS	267, 179
Oxoprolintane[a]	140, 98, 86
Ethamivan-OTMS	223, 294, 193
Methadone[a]	72, 91, 223
Cocaine[a]	82, 182, 105
Levorphanol-OTMS	59, 329, 150
Benzoylecgonine-OTMS	82, 240, 105
Pentazocine-OTMS	289, 110, 244
Dihydrocodeine-OTMS	373, 146, 236
Codeine-OTMS	371, 178, 234
Ethylmorphine-OTMS	385, 192, 146
Morphine-(OTMS)$_2$	429, 236, 146
Phenazocine-OTMS	302, 105, 144

[a] Underivatized form.

TABLE 9.

Doping Class of Androgenic Anabolic Steroids

Androgenic Anabolic Steroids

Bolasterone	Boldenone	Clostebol
Dehydrochlormethyltestosterone		Fluoxymesterone
Mesterolone	Methandienone	Methenolone
Methyltestosterone	Nandrolone	Norethandrolone
Oxandrolone	Oxymesterone	Oxymetholone
Stanozolol	Testosterone	and related substances

Other Anabolic Agents

β-2 agonists (e.g., clenbuterol)

sary to confirm a presumptive positive sample. Therefore, GC/MS with high sensitivity and specificity has been regarded as the most reliable technique so far. In many cases, GC/MS has been used to analyze the steroids either qualitatively or quantitatively.[13]

Anabolic steroids are extensively metabolized; hydroxylation, reduction, oxidation, and conjugation are common metabolic pathways of anabolic steroids in humans. As far as conjugation patterns are concerned, steroids can be divided into two groups, nonconjugated (free) steroid and conjugated steroid. Donike reported that methandienone, oxandrolone, fluoxymesterone, stanozolol, dehydrochlorme thyltestosterone, and formyldienolone can be detected somewhat easily in the nonconjugated fraction. These anabolic steroids have a 17β-hydroxy-17α-methyl group in

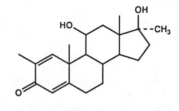

methandienone

oxandronone

fluoxymesterone

stanozolol

dehydrochloromethyl-
testosterone

formyldienolone

Figure 2.
Structure of anabolic steroids. (A) Free fraction; (B) conjugated fraction.

their structure (Figure 2A). The other anabolic steroids (Figure 2B) are excreted mainly as glucuronides.

a. Analysis

A typical example of analysis (see Scheme 1) of anabolic steroids in urine is described as follows:[14]

- *Preparation of XAD-2 column.* A 3-mm diameter glass ball was introduced in the pasteur pipette. The XAD-2 slurry was washed with acetone, methanol, and distilled water and was filled into the column until a bed of 25 mm height was achieved. A final washing with 2 ml of distilled water was carried out before applying the urine.

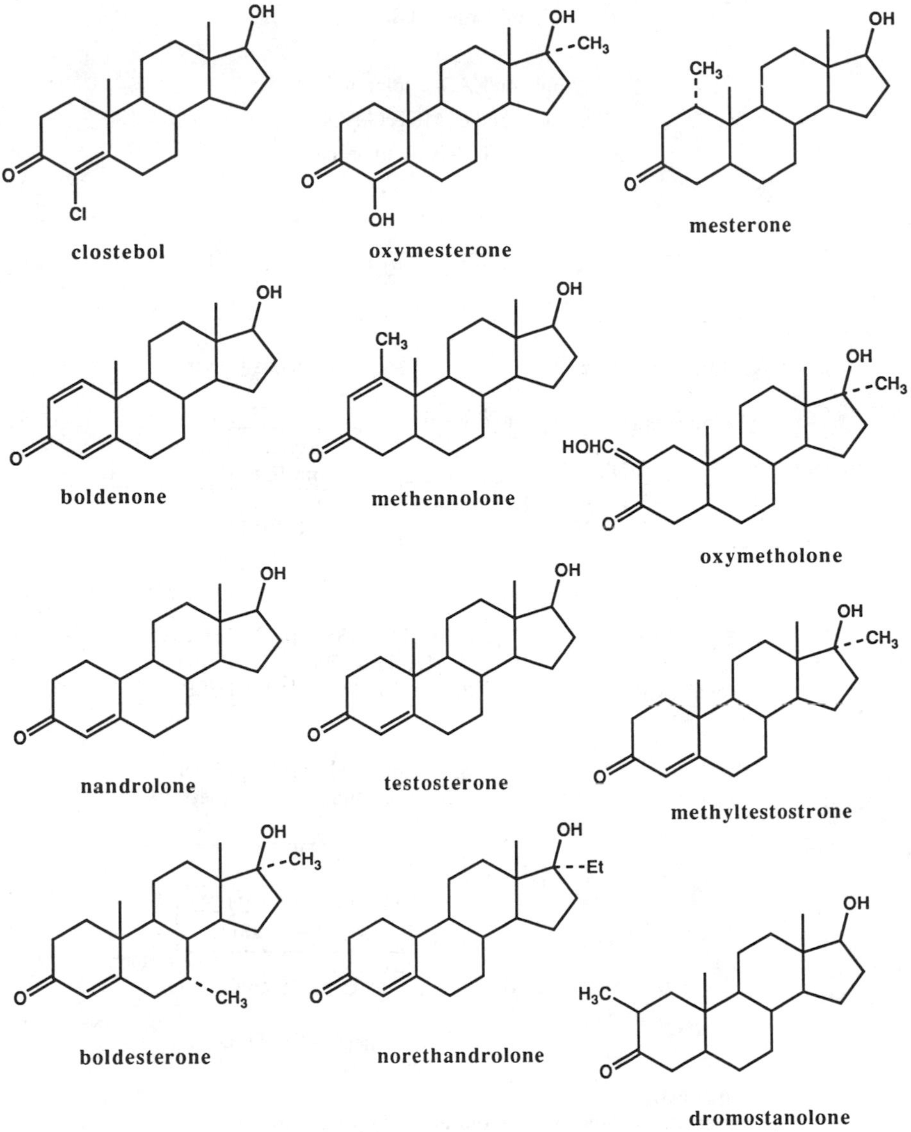

clostebol

oxymesterone

mesterone

boldenone

methennolone

oxymetholone

nandrolone

testosterone

methyltestostrone

boldesterone

norethandrolone

dromostanolone

Figure 2. (continued)

- *Isolation and derivatization of the nonconjugated steroids.* Urine (5 ml) and the internal standard, stanozolol (2 ppm, 25 µl), were applied to the column. The XAD-2 column was washed with the same volume of water. The lipophilic-adsorbed fraction, containing both the free and conjugated steroids, was eluted with 2.7 ml of distilled methanol applied in 0.9 ml fractions. The methanolic extract was brought to dryness

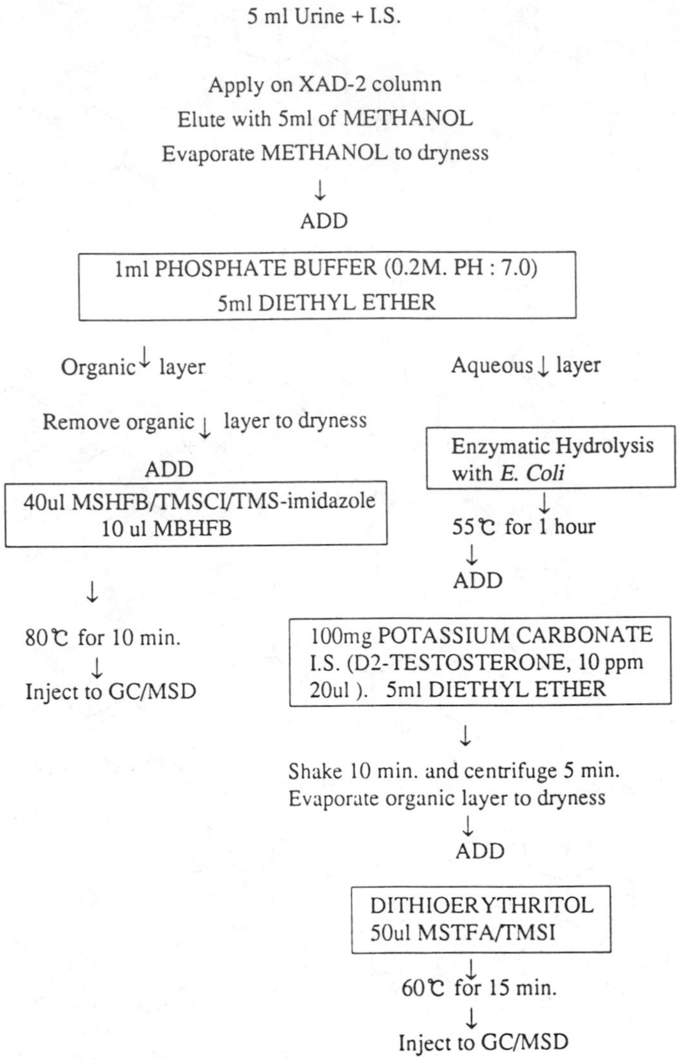

5 ml Urine + I.S.

Apply on XAD-2 column

Elute with 5ml of METHANOL

Evaporate METHANOL to dryness

↓

ADD

```
┌─────────────────────────────────────────────┐
│   1ml PHOSPHATE BUFFER (0.2M. PH : 7.0)       │
│   5ml DIETHYL ETHER                           │
└─────────────────────────────────────────────┘
```

Organic ↓ layer   Aqueous ↓ layer

Remove organic ↓ layer to dryness

```
                                    ┌──────────────────────┐
                                    │ Enzymatic Hydrolysis  │
            ADD                     │ with E. Coli          │
┌───────────────────────────────┐  └──────────────────────┘
│ 40ul MSHFB/TMSCl/TMS-imidazole │           ↓
│ 10 ul MBHFB                    │      55 ℃ for 1 hour
└───────────────────────────────┘           ↓
                                           ADD
```

```
      ↓                    ┌─────────────────────────────────────┐
  80 ℃ for 10 min.         │ 100mg POTASSIUM CARBONATE            │
      ↓                    │ I.S. (D2-TESTOSTERONE, 10 ppm        │
  Inject to GC/MSD         │ 20ul ).  5ml DIETHYL ETHER           │
                           └─────────────────────────────────────┘
```

↓

Shake 10 min. and centrifuge 5 min.

Evaporate organic layer to dryness

↓

ADD

```
┌───────────────────────┐
│ DITHIOERYTHRITOL      │
│ 50ul MSTFA/TMSI       │
└───────────────────────┘
```

↓

60 ℃ for 15 min.

↓

Inject to GC/MSD

Scheme 1.

Experimental steps for steroid analysis procedure.

with a vacuum rotary evaporator. The dried residues were redissolved
in 1 ml of 0.2 M phosphate buffer (pH 7.0) and 5 ml of ether. After
shaking for 5 min and centrifugation at 2500 rpm for 5 min, two layers
were separated. The phosphate buffer layer was subjected to enzymatic
hydrolysis. The ether layer was dried with a rotating evaporator, and
the residue was dried in a vacuum desiccator over P_2O_2-KOH for at least
30 min. The residue was treated with 35 µl of the reagent mixture
MSTFA:Trimethylchloro-silane (TMS-Cl):TMS-imidazole (100:5:2, v/v/

v) and heated to 80°C for 5 min. Then 5 μl of N-methyl-
bisheptafluorobutyramide (MBHFB) was added and the solution was
heated for an additional 10 min.

- *Enzymatic hydrolysis and derivatization of conjugated steroid.* To the phosphate buffer layer, 25 μl of β-glucuronidase were added and the mixture
  was heated at 55°C for 1 h. After cooling, 1,2-dideuterotestosterone (10
  ppm, 20 μl), internal standard for conjugated steroid analysis, 100 mg of
  K_2CO_3, and 5 ml of diethylether were added. After 30 s of shaking in a
  vortex-mixer, 1 g of anhydrous sodium sulfate was added under continuous vortex mixing. After 10 min at room temperature, the tubes
  were centrifuged at 2500 rpm. The ethereal phase was transferred to
  another centrifuge tube and was concentrated by use of a vacuum
  rotary evaporator. Before subjecting the residue to the derivatization
  procedure, it was dried in a desiccator over P_2O_5-KOH for 30 min. To the
  dried residue, 50 μl of MSTFA:Trimethyliodosilane (TMSI) (1000:2) containing 2 mg/ml of dithioerythritol was added and the tubes were
  heated at 60°C for 15 min. To derivatize the hydroxyl groups only, 50 μl
  of MSTFA:TMSCl (100:2) was added and heated at 60°C for 3 min.

- *GC/MS operating conditions for nonconjugated fraction.* A cross-linked 5%
  phenylmethylsilicone capillary column (length 17 m, i.d. 0.2 mm, film
  thickness 0.33 μm) was connected into the ion source. Samples were
  injected in the splitless mode. Temperatures of injector and transfer line
  were set at 300°C. Oven temperature was initially 180°C, ramped by
  25°C/min to 300°C, and held for 3 min. The carrier gas was helium at
  a flow rate of 0.7 ml/min (at 180°C).

- *GC/MS operating conditions for conjugated fraction.* The GC was equipped
  with a 17-m HP-1 fused-silica capillary column, 0.2 mm i.d., 0.11-μm
  film thickness. The oven temperature was programmed from 180 to
  224°C at a rate of 4°C/min and then to 300°C at a rate of 15°C/min. The
  carrier gas was hydrogen with a flow rate of 1.2 ml/min (at 180°C). The
  split ratio was 1:10.

b. Results and Discussion

i. Screening Step

The steroid screening procedure should be able to detect major metabolites and one or two other abundant metabolites. In addition, two to
three characteristic ions for each metabolite should be selected on the basis
of their mass fragmentation. The relative retention times of the chosen
metabolites and the selected ions are listed in Table 10 (free fraction of
anabolic steroid) and Table 11 (conjugated fraction of anabolic steroids
and endogenous steroids).

Also, as shown in Table 10, most anabolic steroids in the free fraction
are metabolized via 6-hydroxylation and all of these metabolites have the
m/z 143 ion in their mass spectra. This ion is produced from the typical

TABLE 10.

SIM Characteristic Ions and Retention Times of Derivatized Main Metabolites of Anabolic Steroids (Free Fraction)

Steroid	Metabolite	Derivative	RRT (min)	Selected ions (m/z)
Fluoxymesterone	6-OH-Fluoxymesterone	6,11,17-tris-O-TMS, 3--enol-TMS	0.799	143, 642, 522, 462
Methandienone	6-OH-Methandienone	6,17-bis-O-TMS	0.085	143, 209, 281, 460
Oxandrolone	—	17-O-TMS	0.915	143, 309, 321, 363
Stanozolol	3'-OH-epistanozolol	N-HFB-3', 17-bis-O-TMS	0.947	143, 669, 684
	3'-OH-Stanozolol	N-HFB-3'-17-bis-O-TMS	1.085	643, 378
IS (stanozolol)	—	N-HFB-O-TMS	1.000	143, 581, 596
Oral-turinabol	6-OH-Oral turinabol	6,17-bis-O-TMS	1.042	143, 315, 317, 243
Formyldienolone	17α-Methylandrosta-1,4-diene-2-hydroxy-methylene-11α, 17 β-diol-3-one	2-(O-TMS-methylene)-11, 17-bis-O-TMS	1.181	143, 367, 457, 562

D-ring cleavage of steroids (Figure 3). It should be noted that formation of the m/z 143 ion is characteristic for trimethylsilylated 17α-methyl-17β-hydroxy-anabolic steroids.

In the case of stanozolol, the main metabolites are 3'-hydroxystanozolol and its epimer.[15] Because stanozolol is metabolized to the glucuronide form exclusively, the concentration of the parent compound in the free fraction is negligible and nondetectable in GC chromatograms. Therefore, stanozolol can be used as an internal standard, that is, a retention time marker for the analysis of anabolic steroids in the free fraction.

Peaks of many endogenous steroids such as androsterone and etiocholanolone, which are produced naturally in the body, and metabolites of ingested substances such as vitamins appear in the chromatogram at retention times very close to those of banned steroids. The peaks of these endogenous steroids may be helpful in interpreting the findings for exogenous steroids. For example, the peak height ratio of testosterone to that of other endogenous androgenic steroids assists in evaluating a case of possible testosterone administration.

The m/z 432 ion, selected for screening of methenolone, mesterolone, and methyltestosterone, is the molecular ion of dehydroepiandrosterone (DHEA), 5α-androstanedione, and epitestosterone, whose peaks appear near the peaks of the metabolites and help to determine their retention times accurately.

TABLE 11.

Relative Retention Times and Selected Ions of O-TMS Derivatives of Conjugated Steroids (or Metabolites) and Endogenous Steroids

Substance	Metabolite (TMS derivative)	RRT	Selected ions
Nandrolone	cis-Norandrosterone	0.6626	405, 420, 422
	Noretiocholanolone	0.7417	405, 420
cis-Androsterone		0.7776	434
Etiocholanolone		0.7936	434
Dromostanolone	3ε-hydroxy-2α-methyl-5α-androstan-17-one	0.8326	448, 433, 343
Methenolone	3ε-hydroxy-1-methylene-5α-androstan-17-one	0.8879	431, 446, 42
DHEA		0.8393	432
Mesterolone	3ε-hydroxyl-1α-methyl-5α-androstan-17-one	0.9202	448, 433, 432
5α-Androstanedione		0.9203	432
Methyltestosterone	5ε-Tetrahydromethyl-testosterone	0.9232	435, 450, 432
Epitestosterone		0.9324	432
Methyltestosterone	5ε-Tetrahydromethyl-testosterone	0.9387	435, 450, 432
Mesterolone	3ε-Hydroxy-1α-methyl-5α-androstan-17-one	0.9728	448, 433
Androstendione		0.9732	430
Boldenone		0.9810	430, 415, 206
D2-Testosterone		1.000	434
Testosterone		1.0016	432
Norethandrolone	5ε-Tetrahydronorethan-drolone	1.0045	421, 157, 331
11β-OH-Androsterone		1.0319	522
11β-OH-Etiocholanolone		1.0539	522
Clostebol	4-Chloro-3ε-hydroxy-5ε-androstan-17-one	1.0582	466, 451, 463
Norethandrolone	5ε-Tetrahydronorethan-drolone	1.0651	421, 157, 331
Bolasterone		1.1749	355, 445, 460
Norethandrolone	Norpregnantriol	1.3549	421, 331, 245
Oxymesterone		1.4631	534, 519, 143
Oxymetholone	3ε, -6ε, -17β-Trihydroxy-2-hydroxymethyl-17α-methyl-androstane	1.5641	550, 495, 143

Because testosterone is one of the endogenous substances, determining a positive doping case of testosterone presents a problem to the laboratory. Under the medical code drawn up by the International Olympic Committee, the definition of a testosterone positive depends upon the administration of testosterone or the use of any other manipulation having the result of increasing the ratio in urine of testosterone/epitestosterone

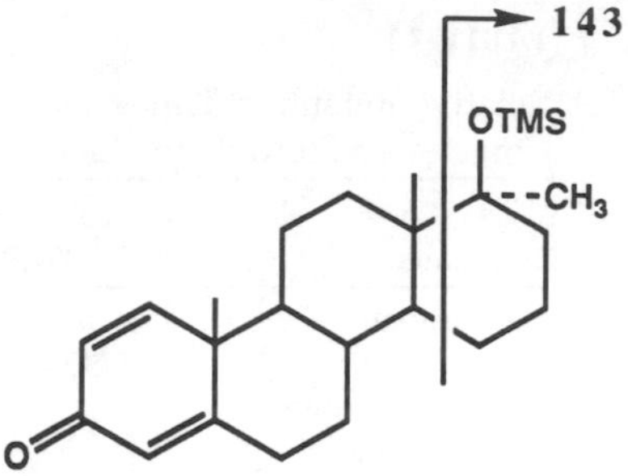

Figure 3.
D-ring cleavage of steroid.

(T/E) to above 6. The T/E ratio can be calculated from the peak height values of testosterone and epitestosterone m/z 432.

In many cases, a nandrolone-positive case requires extra care because of interference from the vitamin E metabolite. The peak of the bis-TMS derivative of α-tocopheronolactone (m/z 422) has nearly the same retention time as that of *cis*-norandrosterone (m/z 422) which is a metabolite of nandrolone. Major peaks from vitamin E metabolites are m/z 422, 405, and 420 which are the same as those of nandrolone metabolites. When the peak of m/z 422 at the retention time of *cis*-norandrosterone is high, it appears as if the nandrolone-free sample were nandrolone positive, but when the peak heights of m/z 405 and 420 are higher than 1% of the peak heights of m/z 422, then it is suspected that the sample might contain nandrolone. Every suspected nandrolone-positive sample should be reextracted and derivatized into mono-TMS compounds and then these compounds can be identified easily.

ii.   Confirmation Step

If a sample is presumptive positive, it should be re-extracted and re-analyzed. In all cases except testosterone, a presumptive positive, a blank, and a positive control sample should be extracted at the same time with the same procedure as the first screening and analyzed by GC/MS.

For the confirmation step, more than five characteristic ions (Table 12) were chosen for ion chromatograms in the SIM mode, and the full mass spectrum of the banned steroid was obtained in the scan mode. In conjugated fraction, if a sample contains α-tocopheronolactone and *cis*-norandrosterone which appear at nearly the same time in the screening procedure, the peaks of their re-analyzed data should be separated. When the more polar capillary columns are used, the peaks of their bis-TMS derivatives are to be separated. Otherwise, the re-extracted sample can be derivatized with the mixture of MSTFA and TMSCI (100:2) which derivatizes only the free hydroxyl group of steroids. Then the peaks of mono-TMS-*cis*-norandrosterone and bis-TMS-α-tocopheronolactone are separated. Presumptive positive testosterone samples are re-extracted three times without an ISTD which can be an interferent. A calibration curve is

TABLE 12.

Selected Ions Chosen for Confirmation of Presumptive
Anabolic Steroids Positive Samples

Substances	Selected ions
Bolasterone	464,[a] 460,[b] 449, 445, 355, 143
Boldenone	432,[a] 430,[b] 417, 415, 325, 229, 206, 194, 191
Clostebol	468,[a] 466,[b] 453, 451, 431, 363, 361
Dromostanolone	448,[a] 433, 358, 343, 365, 182, 169
Mesterolone	448,[a] 433, 358, 343, 270, 143
Methenolone	466,[a] 431, 251, 195, 169
Methyltestosterone	450,[a] 435, 360, 365, 255, 143
Nandrolone	422, 420,[a] 405, 315, 225
	422, 348,[a] 333, 258 (mono-TMS)
Norethandrolone	466,[a] 421, 375, 331, 287, 254, 241, 157
Oxymesterone	534,[a] 519, 444, 429, 389, 358, 269, 229, 143
Oxymetholone	640,[a] 625, 550, 495, 460, 370, 143
Testosterone	434, 432, 430, 417, 522

[a] Molecular ion of TMS derivative of main metabolite.
[b] Molecular ion of TMS derivative of parent steroid.

prepared in order to quantitate testosterone. The range of T/E ratio used in the calibration curve is 4 to 10. The T/E ratios from the three re-analyzed samples are averaged, and the mean value is then corrected by multiplying by the relative response factor.

D. Analysis of Diuretics

Diuretics are widely used therapeutically for hypertension and edema resulting from cardiac or renal failure. These agents typically increase the urine flow and the net renal excretion of both sodium and water. Diuretics are classified in different groups according to their pharmacological properties: carbonic anhydrase inhibitors, loop diuretics, thiazides, potassium-sparing diuretics, and others. These drugs are generally excreted in unchanged form to a high extent.

Diuretics are sometimes misused by competitors for two main reasons, namely to reduce weight quickly in sports where weight categories are involved and to reduce the concentration of drugs in urine by producing a more rapid excretion of urine to attempt to minimize detection of drug misuse. Rapid reduction of weight in sports cannot be justified medically. Health risks are involved in such misuse because of serious side effects which might occur.

Furthermore, deliberate attempts to reduce weight artificially in order to compete in lower weight classes or to dilute urine constitute clear manipulations which are unacceptable on ethical grounds. For these

TABLE 13.

Doping Class of Diuretics

Acetazolamide	Amiloride	Bendroflumethiazide
Benzthiazide	Bumetanide	Canrenone
Chlormerodrin	Chlorthalidone	Dichlorphenamide
Ethacrynic acid	Furosemide	Hydrochlorothiazide
Mersalyl	Spironolactone	Triamterene
and related compounds		

reasons, the IOC has decided to include diuretics on its list of banned classes of drugs. Table 13 represents examples of diuretics banned by IOC.

1. Analysis of Diuretics

There are several chromatographic methods reported for the separation, detection, and quantitative measurement of individual diuretic agents in the biological fluids. Published methods included those based on thin-layer chromatography,[16] gas-liquid chromatography,[17] GC/MS,[18] and high-performance liquid chromatography.[19]

GC methods have been based on alkylation of the sulfonamide group and other groups containing O- or N-bonded hydrogen atoms of the diuretics, and high sensitivity has been obtained with electron capture detection or nitrogen selective detection.

LC systems have utilized reversed-phase C_{18} stationary phases and mobile phases with acetate or phosphate buffer and acetonitrile or methanol as an organic modifier. UV detectors or fluorescence detectors have been selected according to the property of the drug.

Many procedures for sample preparation for reversed-phase HPLC have been reported using single extraction by organic solvents or single column extraction. Sample preparation steps include derivatization or hydrolyzation to enhance sensitivity by pre- or postcolumn techniques.

a. Analysis

A typical example of analysis of diuretics in urine is described as follows:[20]

- *Liquid-liquid extraction (LLE).* A 3-ml urine sample was pipetted into a 15-ml centrifuge tube. Next, 15 µl of IS stock solution (1000 ppm of ethyltheophylline) was added and pH was adjusted with 0.1 g solid phosphate buffer (pH 5, KH_2PO_4; pH 7, KH_2PO_4/K_3PO_4 [1:1]; pH 9, K_2HPO_4; pH 11, K_2HPO_4/K_3PO_4 [3:1]). Then 0.5 g of anhydrous sodium sulfate was added, the solution was vortex mixed, and 5 ml of diethylether was added. The tube was shaken mechanically for 10 min and centrifuged at 2500 g for 5 min. The organic phase was transferred to a second tube and evaporated to dryness under a vacuum rotary evaporator. The dried residue was reconstituted with 200 μl of methanol and filtered through the sample clarification filter (pore size, 0.45 μm), and 5 μl was injected into the HPLC.

- *Derivatization.* Sample extract as described above was evaporated to dryness. The residue was dissolved in 200 μl of acetone and 20 μl of methyliodide, 100 mg of potassium carbonate was added, and the solution was heated in a heating block at 60°C for 2 h. A 2-μl aliquot of methylated product was injected into the GC/MSD.

- *Gas chromatography/mass spectrometry.* A HP fused-silica capillary column coated with cross-linked methylsilicone (SE-30, 16 m × 0.2 mm i.d., 0.33-μm film thickness) was used. The GC temperature program was as follows: initial temperature was 200°C, increased by 15°C/min to 280°C, by 10°C/min to 300°C, held for 1.20 min, increased by 20°C/min to 310°C, and held for 2.0 min. Injector temperature was 290°C, transfer line temperature was 290°C, and ion source was 200°C. The injector was operated in the split mode (1:10). The carrier gas was helium at 0.90 ml/min. The mass spectrometer was operated at 70 eV in the electron impact (EI) mode using scan or selected ion monitoring (SIM).

b. Results and Discussion

Generally because the diuretic agents contain sulfonamide, amide, hydroxyl, carboxyl, or amine groups (polar and nonvolatile), poor GC response due to the interaction or adsorption with column materials is observed. It is necessary to derivatize the parent drug to improve the volatility and thermal stability. Under these experimental conditions, nine diuretics (acetazolamide, bendroflumethiazide, bumetanide, chlorthalidone, dichlorphenamide, ethacrynic acid, furosemide, hydrochlorothiazide, and triamterene) were methylated at the position of the *O*- or *N*-bonded hydrogen atom but not in the secondary amine position adjacent directly to an aromatic ring (furosemide and bumetanide). Acetazolamide produced two different methyl derivatives. Canrenone and spironolactone have no polar functional groups for methylation, and can be detected without derivatization. Amiloride and benzthiazide were not methylated.

All nine diuretics were well resolved in 11 min under above conditions, except for canrenone and spironolactone.

One of the two peaks of methylated acetazolamide is characterized by a base peak at m/z 249 (M^+–CH_3), molecular ion (M^+, 264), and m/z 108 $[SO_2N(CH_3)_2]^+$. The other methyl derivative yields a base peak at m/z 43 due to CH_3 CO and m/z 222 from the loss of ketene (M^+–CH_2CO).

Canrenone was eluted without derivatization and a base peak of m/z 267 and an intense molecular ion (M^+, 340) were observed. Spironolactone and canrenone had the same retention time and mass spectrum, seemingly due to the loss of thioacetyl ($SOCCH_3$) on the injection port or on the column.

There are five diuretics (chlorthalidone, furosemide, hydrochloro-thiazide, dichlorphenamide, and ethacrynic acid) that contain one or two chlorine atoms. Their mass fragment ions (m) are accompanied by (m + 2) ion for a one Cl atom-containing compound, and by (m + 2) and (m + 4) ion for two Cl atoms because of the isotopic effect of chlorine. For example, the methyl derivative of chlorthalidone yields a base peak m/z 363 from the loss of a methoxy radical, and m/z 287 ($M^+-SO_2N(CH_3)_2$), and m/z 255 ($363-SO_2N(CH_3)_2$). Those are followed by m/z 365, 289, and 257. The methyl derivative of ethacrynic acid with two chlorine atoms shows a low intensity molecular ion (M^+, 316). A base peak at m/z 261 is from the loss of the butenyl radical (C_4H_7) from the molecular ion and m/z 243 is from the cleavage of aromatic ether bond ($M^+-CH_2COOCH_3$). These fragments are accompanied by three isotopic peaks, m/z 318 and 320, m/z 263 and 265, and m/z 245 and 247.

The detection limits of methylated diuretics were ranged over 20 ng as an absolute amount in the scan mode. The selected ion monitoring mode may be used to improve the detection limits. The detection limits by selected ion monitoring mode using only the base peak of each derivative were 0.05 ng, except those of acetazolamide, canrenone, ethacrynic acid, and hydrochlorothiazide, which were 0.2 ng at a signal-to-noise ratio of 3. This sensitivity could be increased by adjusting the urine volume and by using the splitless mode.

The mass fragmental pathways for methyl diuretic agents have been demonstrated with the use of methyl isotope mass spectra.[21] The prominent ions of the methyl and methyl-d_3 derivatives of each diuretic agent are listed in Table 14.

E. Analysis of Corticosteroids

Both naturally occurring and synthetic corticosteroids are used primarily as anti-inflammatory drugs that also relieve pain. They affect the blood level of natural corticosteroids.

It was known that athletes in sports such as cycling and weightlifting used corticosteroids to improve performance. Since 1975, the IOC Medical Commission has restricted such use except for legitimate medical purposes.

Chemical structures common to corticosteroids included Δ^4, 3-keto, 20-keto, and the 21-hydroxy group (Figure 4). When administered to humans, corticosteroids are bound to protein. Their major metabolites are glucuronides, and they are excreted as a conjugated form or parent form in urine.

1. Analysis of Corticosteroids

Analyses based on radioimmunoassay procedures are characterized by high sensitivity, but this benefit may be offset by the lack of specificity

TABLE 14.

Characteristic Ions of Methyl and Methyl-d$_3$ Derivatives of Ten Diuretic Agents

Diuretics		Mol wt	Base peak (m/z)	Other ions (m/z)
Acetazolamide	CH$_3$[a]	264	249	43, 83, 108, 264
	CD$_3$[b]	273	258	46, 86, 114, 273
Bendroflumethiazide	CH$_3$	477	386	42, 91, 278
	CD$_3$	489	398	45, 91, 284
Bumetanide	CH$_3$	406	254	318, 363, 406
	CD$_3$	415	257	321, 372, 415
Chlorthalidone	CH$_3$	394	176	255, 257, 287, 289, 363, 365
	CD$_3$	406	182	258, 260, 294, 296, 372, 374
Dichlorphenamide	CH$_3$	360	44	108, 144, 253, 255, 360, 362
	CD$_3$	372	50	114, 144, 260, 262, 372, 374
Ethacrynic acid	CH$_3$	316	261	243, 245, 263, 316, 318
	CD$_3$	319	264	243, 245, 266, 319, 321
Furosemide	CH$_3$	372	81	53, 96, 372, 374
	CD$_3$	381	81	53, 96, 381, 383
Hydrochlorothiazide	CH$_3$	353	310	218, 220, 288, 290, 312, 353, 357
	CD$_3$	365	319	221, 223, 300, 302, 321, 365, 367
Mefruside	CH$_3$	410	85	43
	CD$_3$	416	85	43
Triamterene	CH$_3$	337	336	279, 294, 307, 322, 337
	CD$_3$	355	355	289, 307, 323, 337, 354

[a]   Methyl derivative.
[b]   Deuterated methyl derivative.

due to cross-reactivity of related compounds.[22] Analyses of corticosteroids by GC and GC/MS methods,[23] while highly sensitive and specific, required derivatization prior to injection because the compounds are thermally labile and their volatility is too low for direct GC analysis.

Recently, several HPLC procedures were reported for the detection and assay of synthetic or natural corticosteroids in biological fluids.[24] Because there is no possibility of thermal decomposition, HPLC offers the advantage that the corticosteroids can be analyzed without derivatization. HPLC interfaced to a mass spectrometer (thermospray LC/MS) offers the further advantage of highly specific detection of these steroids in biological extracts.

a.   Analysis

A typical example of analysis of corticosteroids in urine is described as follows:[25]

- *Liquid-liquid extraction (LLE).* A 3-ml urine samples was pipetted into a 15-ml centrifuge tube. Next, 15 µl of IS stock solution (100 ppm of

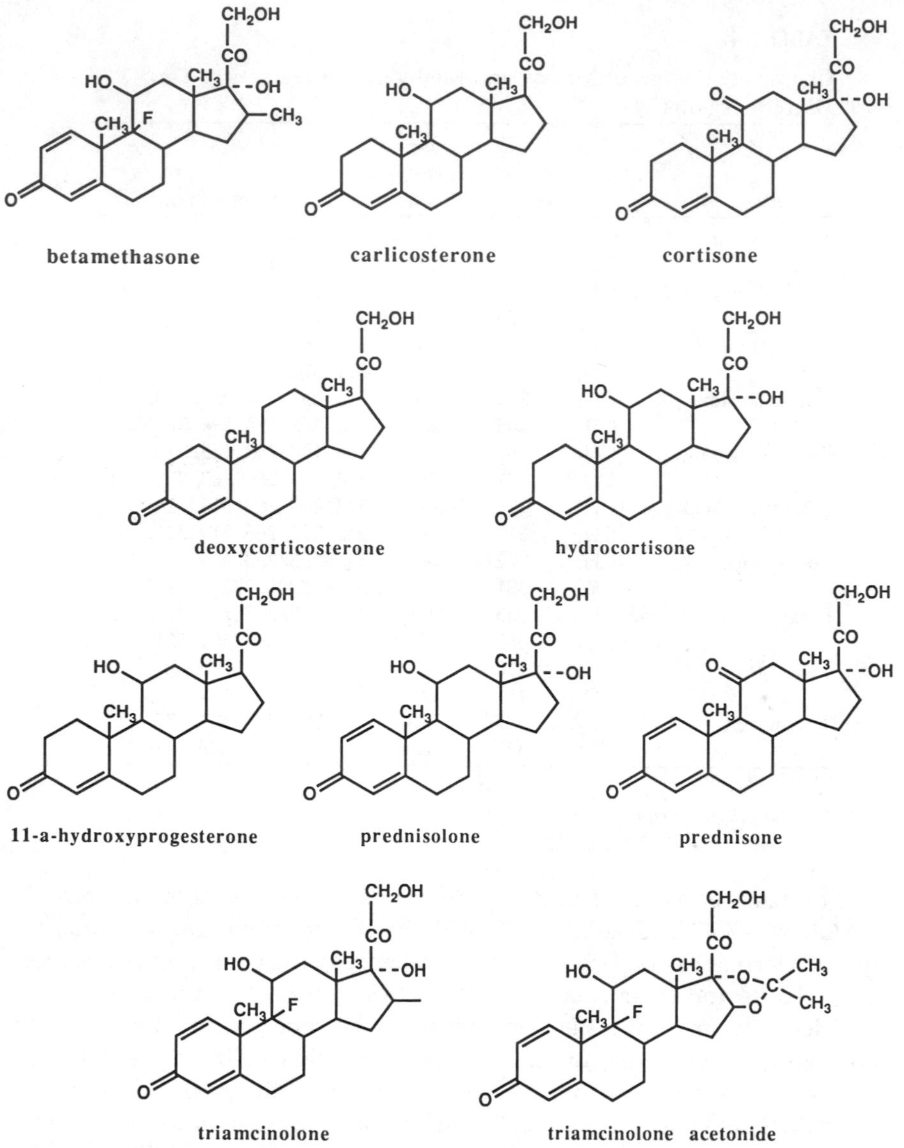

Figure 4.
Chemical structure of corticosteroids.

betamethasone) was added and pH was adjusted with 0.1 g solid phosphate buffer pH 9, K_2HPO_4. Then 0.5 g of anhydrous sodium sulfate was added, the solution was vortex mixed, and 5 ml of diethyl ether was added. The tube was shaken mechanically for 10 min and centrifuged at 2500 g for 5 min. The organic phase was transferred to a second tube and evaporated to dryness under a vacuum rotary evaporator. The dried residue was reconstituted with 200 μl of methanol and filtered through the sample clarification filter (pore size, 0.45 μm), and 5 μl was injected into the HPLC/MS.

- *Thermospray LC/MS.* The LC/MS system used was a HPLC linked via capillary tubing and thermospray vaporizer probe to a mass spectrometer. The MS was tuned daily with polypropylene glycol solution. The LC column was a Hypersil-ODS (60 × 4.6 mm i.d., 3 μm), and the mobile phase was 0.15 M ammonium acetate and methanol with a flow rate of 0.8 ml/min. The solvent composition was as follows: initial methanol was 40%, increased to 50% for 6 min, held for 1 min, then increased to 60% for 3 min and held for 5 min. The optimum vaporizer probe temperature was programmed according to the composition of the mobile phase. The typical probe temperature program was as follows: initial stem temperature was 92°C, decreased to 89°C at the rate of 0.5°C/min, held for 1 min, decreased to 86°C at the rate of 1°C/min, and held for 5 min. Other MS parameters were as follows: ion source temperature, 276°C; emission current, 150 μA; electron energy, 955 eV. The ion source was operated in the positive ion mode with filament-on mode.

b. Results and Discussion

Under the above LC/MS conditions, the total ion chromatogram and mass spectrum of each compound are shown in Figure 5. The fragment or adduct ions of corticosteroids are listed in Table 15.

Protonated molecular ion species $[MH]^+$ were found in all corticosteroids commonly having 3- and 17-keto groups. The base peak $[MH]^+$ was found in compounds containing no 17-hydroxy group. Corticosteroids containing a 17-hydroxy group such as betamethasone, cortisone, hydrocortisone, prednisolone, and prednisone produced a base peak $[MH-60]^+$ from the bond-cleavage between 17- and 20-carbon, and $[MH-30]^+$ due to the loss of CH_2O. In addition $[MH-18]^+$ from the loss of H_2O was observed. An ammonium adduct ion $[MNH_4]^+$ was found in corticosteroids with no double bond between 1- and 2-carbon, except for triamcinolone acetonide. Both triamcinolone and triamcinolone acetonide showed m/z 359 and m/z 377 due to the side chain cleavage between the 15- and 16-carbon.

The retention time and characteristic mass fragment ions $[MH^+-60]$, $[MH^+-30]$, and $[MH^+-18]$, as well as the protonated ion $[MH]^+$ and ammonium adduct ion $[MNH_4]^+$, could be used for the confirmation of each corticosteroid. The detection limits (by scan mode) were 10 ng for 11-α-hydroxyprogesterone, prednisolone, prednisone, and triamcinolone; 30 ng for cortisone and hydrocortisone; and 50 ng for the other corticosteroids tested. Sensitivity could be increased by using SIM mode with two or three ions producing the highest intensity; the detection limit was thus lowered to approximately 1 to 5 ng for each corticosteroid.

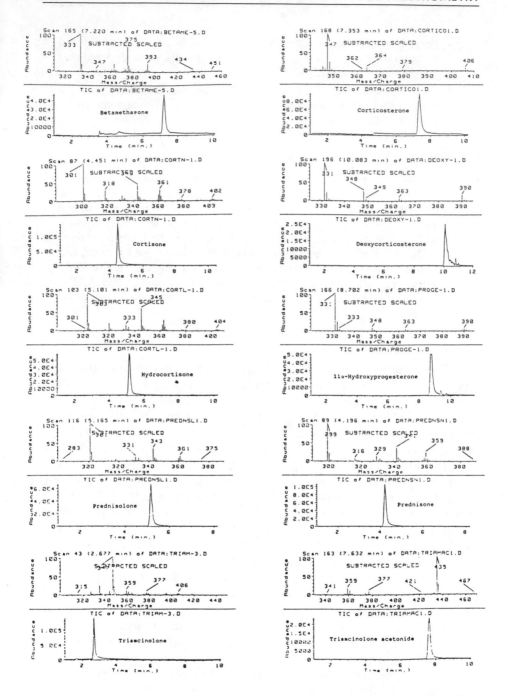

Figure 5.

Total ion chromatograms and mass spectra of corticosteroids.

TABLE 15.

Mass Fragment Ions of Corticosteroids

Corticosteroids	Molecular weight	Base peak	MH+	MNH$_4^+$	MH+-60	MH+-30	MH+-18 (m/z)
Betamethasone	392	333	393	—	333	363	375
Corticosterone	346	347	347	364	—	—	—
Cortisone	360	301	361	378	301	331	343
Deoxycortico-sterone	330	331	331	348	—	—	—
Hydrocortisone	362	303	363	380	303	333	345
11-α-OH-Progesterone	330	331	331	348	—	—	—
Prednisolone	360	301	361	—	301	331	343
Prednisone	358	299	359	—	299	329	341
Triamcinolone	394	347	395	—	335	365	377
Triamcinolone acetonide	434	435	435	452	—	—	417

F. Analysis of Beta-Blockers

There is now a wide range of effective alternative preparations available in order to control hypertension, cardiac arrhythmias, angina pectoris, and migraine in addition to beta-blockering drugs. Due to the continued misuse of beta-blockers in some sports where physical activity is of no or little importance, beta-blockers are now routinely tested in these sports, i.e., shooting, archery, and others. It is evident that beta-blockers testing will be done in endurance events which necessitate prolonged periods of high cardiac output and large stores of metabolic substrates in which beta-blockers would severely decrease performance capacity. Table 16 represents examples of beta-blockers banned by IOC.

1. Analysis of Beta-Blockers

a. Analysis

Beta-blockers can be analyzed by the same method as narcotics analysis, described in Section II.B.2.

b. Results and Discussion

The functionality and polarity of beta-blockers are very similar to those of phenolalkylamines. The NTFA-OTMS derivatives of beta-blockers were easily prepared by trimethylsilylation with MSTFA followed by selective N-trifluoroacylation with MBTFA. Acebutolol, alprenolol, atenolol, metoprolol, and propranolol can be considered as derivatives of oxypropanolamine having the general structure (I). Other beta-blockers, including sotalol and labetalol, are derivatives of phenylethylamine (II):

TABLE 16.

Doping Class of Beta-Blockers

Classes of Drugs Subject to Certain Restrictions

Beta-blockers, e.g.

Acebutolol	Alprenolol	Atenolol
Labetalol	Metoprolol	Nadolol
Oxprenolol	Propranolol	Sotalol
and related compounds		

$$\begin{array}{c} OH \\ | \\ R_1\text{–O–}CH_2\text{–CH-}CH_2\text{–NH}(CH_3)_2 \end{array} \qquad (I)$$

$$\begin{array}{c} OH \\ | \\ R_1\text{–CH–}CH_2\text{–NH–}R_2 \end{array} \qquad (II)$$

R_1 is usually a substituted aromatic ring and R_2 is an isopropyl or *tert*-butyl group.

In the mass spectra of the NTFA-OTMS derivatives of oxypropanolamines, the base peak at m/z 284 is a result of the loss of the aryloxy group from the molecular ion. The m/z 129 ion arises from the 2-propanol fragment. The m/z 284 and 129 ions characterized the oxypropanolamines containing an isopropylamine group.

Fragmentation of the phenylethylamine series (II) takes place alpha to the nitrogen atom to yield the corresponding benzylic fragment. Mass spectra of the sotalol and labetalol derivatives showed the corresponding base peaks at m/z 344 and 292, respectively. The amine function containing a *tert*-butyl group on the nitrogen atom in nadolol was not trifluoroacylated with MBTFA.

The bulky group on the nitrogen atom hinders the introduction of a TFA group onto the nitrogen atom and onto the amino group. In the mass spectrum of the nadolol derivative, the base peak at m/z 86 is a result of α-fission of the major side chain. The characteristic ions of nine beta-blockers are shown in Table 17. Screening of the athletes' samples was performed on a GC/MS using selected ion monitoring. The ions monitored for screening of the oxypropanolamine series were m/z 284 and 129; for labetalol they were m/z 292 and 316; for sotalol, m/z 344 and 272; and for nadolol, m/z 86 and 510.

TABLE 17.

Characteristic Ions of Derivatives of Nine Beta-Blockers

Drug	Mol wt	Base peak	Characteristic ions (m/z)
Acebutolol-NTFA-OTMS	504	284	129, 278, 504
Alprenolol-NTFA-OTMS	417	284	129, 159, 402
Atenolol-NTFA-OTMS	434	284	158, 129, 228
Metoprolol-NTFA-OTMS	435	284	129, 235, 420
Oxprenolol-NTFA-OTMS	432	284	129, 166, 418
Propranolol-NTFA-OTMS	427	284	129, 169, 412
Nadolol-(OTMS)$_2$	525	86	147, 510, 409
Labetalol-NTFA-(OTMS)$_2$[a]	550	292	91, 316, 535
Sotalol-NTFA-NTMS-OTMS	512	344	272, 281, 497

[a] An amide function of labetalol-NTFA-(OTMS)$_2$ was dehydrated.

REFERENCES

1a. Prokop, L., *Proceedings of the International Symposium on Drug Abuse in Sports (Doping)*, Park, J., Ed., Doping Control Center, KIST, Seoul, Korea, 1988.

1b. Goldman, B., *Death in the Locker Room, Steroids in Sports*, BL Publisher, Tucson, Arizona, 1984.

2. Dirix, A., Dope control at the Olympic games, in *Proceedings of the International Symposium on Drug Abuse in Sports (Doping)*, Park, J., Ed., Doping Control Center, KIST, Seoul, Korea, 1988, 12.

3. Catlin, D. H., Kammerer, R. C., Hatton, C. K., Sereka, M. H., and Merdink, J. L., *Clin. Chem.*, 33, 319, 1987.

4. Park, J., Park, S. J., Lho, D. S., Choo, H., Chung, B., Yoon, C., Min, H., and Choi, M. J., *J. Anal. Toxicol.*, 14, 66, 1990.

5. Donike, M. and Stratmann, D., *Chromatographia*, 7, 182, 1974.

6. Lho, D. S., Shin, H. S., Kang, B. K., and Park, J., *J. Anal. Toxicol.*, 14, 73, 1990, and references cited in it.

7. Beckett, A. H., Tucker, G. T., and Moffat, A. C., *J. Pharm. Pharmacol.*, 19, 273, 1967.

8. Hartley, R., Smith, I. J., and Cookman, J. R., *J. Chromatogr.*, 342, 105, 1985.

9. Dugal, R., Masse, R., Sanchez, G., and Bertrand, M., *J. Anal. Toxicol.*, 4, 1, 1980.

10. Lho, D. S., Hong, J. K., Paek, H. K., Lee, J. A., and Park, J. S., *J. Anal. Toxicol.*, 14, 77, 1990, and references cited in it.

11. Hara, S. and Hayashi, S., *J. Chromatogr.*, 142, 689, 1987.

12. Hampl, R. and Starka, L., *J. Steroid Biochem.*, 11(1C), 933, 1979.

13. Donike, M., Zimmermann, J., Barwald, K. R., Schanzer, W., Christ, V., Klostermann, K., and Opfermann, G., *Dtsch. Z. Sportmedizin.*, 35, 14, 1984.

14. Chung, B. C., P. Choo, H. Y., Kim, T. W., Eom, K. D., Kwon, O. S., Suh, J. W., Yang, J. S., and Park, J. S., *J. Anal. Toxicol.*, 14, 91, 1990, and references cited in it.

15. Choo, H. Y., Kwon, O. S., and Park, J. S., *J. Anal. Toxicol.*, 14, 109, 1990.

16. Hadzija, B. W. and Mattocks, A. M., *J. Chromatogr.*, 229, 425, 1982.

17. Ervik, M. and Gustavii, K., *Anal. Chem.*, 46, 39, 1974.

18. Fleuren, H. L. J. M. and Van Rossum, J. M., *J. Chromatogr.*, 152, 41, 1978.
19. Hessey, G. A., II, Constanzer, M. L., and Bayne, W. F., *J. Chromatogr.*, 380, 450, 1986.
20. Park, S. J., Pyo, H. S., Kim, Y. J., Kim, M. S., and Park, J. S., *J. Anal. Toxicol.*, 14, 84, 1990, and references cited in it.
21. Yoon, C. N., Lee, T. H., and Park, J. S., *J. Anal. Toxicol.*, 14, 96, 1990.
22. Kuhn, R. W. and Deyman, M. E., *J. Chromatogr.*, 421, 123, 1987.
23. Vrbanac, J. J., Sweeley, C. C., and Pinkstone, J. D., *Biomed. Mass Spectrom.*, 10, 155, 1983.
24. Prasad, V. K., Ho, B., and Haneke, C., *J. Chromatogr.*, 378, 305, 1986.
25. Park, S. J., Kim, Y. J., Pyo, H. S., and Park, J. S., *J. Anal. Toxicol.*, 14, 102, 1990 and references cited in it.

Chapter **4**

Analysis of Accelerants in Fire Debris by Gas Chromatography/Mass Spectrometry

Wolfgang Bertsch and Gunther Holzer

CONTENTS

0-8493-8252-1/95/$0.00+$.50
© 1995 by CRC Press Inc.

I. ABSTRACT

The current status of accelerant analysis by GC/MS is critically assessed. The analytical methodology of laboratories involved in the analysis of suspect arson debris samples is statistically evaluated. Sample preparation procedures are briefly discussed, along with general instrumental methods of analysis. The main body of the chapter deals with GC/MS, as applied to petroleum-based fluids. The impact of interferences from synthetic polymers is illustrated with several examples. The reader is led though the interpretation of several cases involving the use of various accelerants. The need for quality control is emphasized. Approaches are shown to speed up data reduction, eventually leading to side-by-side comparison of the ion chromatograms of the sample to those of an accelerant library. Finally, aspects of automated analysis are introduced in the form of a GC/MS based expert system.

II. INTRODUCTION

A. Background

Arson is a major problem in all industrialized countries. A significant body of literature exists on all aspects of arson, ranging from legal deliberations to technical questions. Arson is a complex subject that requires information from many different perspectives. Several textbooks are available that treat arson in a comprehensive manner, including the topic of chemical analysis.[1-6] Subject areas dealing with particular aspects of arson are often found in review articles. While many journals are published dealing with various aspects of forensic science, none exists in the specific area of the analysis for accelerants in suspect arson cases, often referred to as arson analysis. This chapter focuses on a small but significant aspect of the overall investigation, the chemical analysis of fire debris for the presence of accelerants. While the laboratory results are important in establishing the possibility of an incendiary fire, this information can only be one of many factors to establish the incidence of arson.

When arson is suspected, evidence must be secured for possible prosecution. The rules for collection of evidence are stringent and the investigator must maintain a strict chain of custody. It is very important to begin the investigation as soon as the fire has been extinguished. Volatiles from an accelerant can evaporate if the debris is not secured quickly and placed in sealed containers. Environmental factors such as temperature and moisture, air flow over the debris, and the volatility of the accelerant affect the rate at which residual accelerant evaporates. The investigator must be careful to prevent accidental contamination at the fire scene. Traces of accelerant may be transferred within a site if one is not constantly aware of potential contamination. Contamination may also occur during transport and storage if the sample containers are not properly sealed.

Successful arson investigation is as much a science as it is an art. It takes time and patience. Competent arson investigators are always aware of pitfalls and often develop a remarkable instinct for arson-related fires. Some investigators claim that they can smell petroleum-based liquids. It may well be that some individuals have a highly developed sense of smell. Nevertheless, a nose can be deceived and it is always difficult to remain completely objective. It is necessary to use a scientific approach in establishing the presence of materials at a fire scene that constitute accelerants.

B. Detection of Accelerants at the Fire Scene

The oldest and often most relied upon instrument for on-site detection of accelerants is the human nose. A nose can be a sensitive and selective detector, but it has some disadvantages. Threshold levels of detection are strongly dependent on the chemical nature of the volatile that is sniffed. Most petroleum-based fluids do not produce particularly characteristic odors. Relatively high concentrations must be present for detection by the nose. Its most serious handicap comes from the limited ability of the olfactory senses to discriminate against interferences. Fatigue presents another problem. Prolonged exposure to vapors at the fire scene eventually "dulls" the senses. The nose can thus easily be deceived.

A variety of instruments are available for on-site detection of accelerants. These devices measure some chemical or physical property of the volatiles at the fire scene. The accelerants most frequently used are petroleum-based fluids that consist of volatile components which are usually detected in the gas phase. The most widely applied group of on-site detectors are hand-held devices, more commonly known as "sniffers". Sniffers vary greatly in sophistication and cost. Advanced devices rely on catalytic oxidation, semiconductors, photoionization, fluorescence, and flame ionization. Solid state sensors, especially those based on silicon chip technology, are perhaps the most promising development. Chemiresistors are based on surface acoustic waves generated on a small piezoelectric crystal. Depending on coating and surface, instantaneous detection is possible. Even portable gas chromatographs have been evaluated for field work.[7] While portable detectors are becoming increasingly sophisticated, they generally suffer from a lack of sensitivity and, more importantly, selectivity.

In principle, analytical instruments can be carried to the fire scene for on-site analysis instead of bringing samples to a central analytical laboratory. There are some advantages to this approach. A new generation of portable gas chromatographs which can accomplish on-site analysis has come onto the market. An example is a recently introduced miniature gas chromatograph based on silicon chip technology.[8] This type of instrument signals an interesting development, but direct analysis of fire debris in the field is hampered because of problems with the enrichment of volatiles on the spot. There are also portable instruments that take advantage of the selectivity and sensitivity of mass spectrometry. These are discussed at the end of this chapter.

C. Manipulation and Preservation of Samples at the Fire Scene

Once the fire investigator suspects that a fire is incendiary in nature, samples are stored for chemical analysis of accelerants. The fire scene is photographed before and after samples are taken. Fire debris must be packaged so that volatiles will not escape and contaminants are not introduced into the sample. The most common packing medium used is an unlined, unused metal paint can. Paint cans are virtually air tight and vary in volume from about 1 pt to 1 gal. Cans that are untreated internally may rust with time, especially with corrosive samples. Glass jars can also be used, but they are bulky and have the added disadvantage of fragility.

Plastic bags are also frequently used. They are convenient and do not occupy much space in storage. They have some disadvantages. Common plastics are organic in nature and may be transparent to hydrocarbons. Polyethylene containers are partially permeable to gasoline, leading to the loss of volatiles or contamination from outside sources.[9] Polyester/polyolefin bags are said to be more suitable for packaging of fire debris.[10] Bags manufactured by the Kapak Corporation are frequently used for arson evidence packaging, although some loss of polar substances, such as methanol, may occur. A recent report pointed out problems with some batches of the Kapak bags.[11] It appears that changes in processing of the material at a new manufacturing facility inadvertently introduced some contamination. Dietz et al. reported the presence of a hydrocarbon background in some bags manufactured after 1985. This incident points out the importance of quality control. Periodic background and equipment checks prevent such potential artifacts.

D. The Role of the Forensic Chemist

Most chemists who examine fire debris are not experts in fire investigation, yet they are frequently asked how materials react in fires. Unless specially trained, the chemist should stay away from speculation and only report on materials and substances that may have accelerated a fire. The chemist should work without bias and should not be told details about the fire scene and the parties involved.

Appearance in a court of law is one of the responsibilities of a forensic chemist. There are some peculiarities that apply only to expert witnesses testifying in scientific fields.[12,13] Most likely, there will be a need to communicate the findings on a basic, nontechnical level. This may be as simple as discussing sample preparation procedures. On the other hand, it may also be necessary to introduce principles of gas chromatography/pattern recognition. With modern equipment such as GC/MS, it may be necessary to make the court familiar with principles of mass spectrometry and data processing. To facilitate transfer of information some simple diagrams and exhibits should be prepared.[14]

III. A BRIEF SURVEY OF CURRENT ANALYTICAL METHODS AND PROCEDURES

Laboratory verification is currently not necessary. This is quite different from some other laboratories, i.e., those following EPA protocols and clinical laboratories. The community of forensic scientists who are en-

TABLE 1.

Sample Types and Matrixes Used in the Collaborative Testing
Program, 1988–1992

Year	Description of accelerant	Matrix	ASTM class
1988	Lacquer thinner	Cotton	0
	95% Evaporate gasoline	Cotton	2
	Gasoline/diesel mixture	Cotton	2 + 5
	Medium petroleum distillate	Cotton	3
1989	Gasoline	Uncharred carpet	2
	Weathered gasoline	Uncharred carpet	2
	Gasoline/fuel oil #2 mixture	Uncharred carpet	2 + 5
1990	Coleman lantern fluid	Charred carpet	1
	Coleman lantern fluid	Uncharred carpet	1
1991	Gasoline	Polystyrene (napalm)	2
	Gasoline/motor oil	None (neat liquid)	2 + 5
1992	Turpentine	Oak parquet	0
	Turpentine	Yellow pine	0

Note: Compiled from annual reports, Collaborative Testing Services,
Herndon, VA.

gaged in fire debris analysis has set up a voluntary system of self-testing
and evaluation. Laboratory analysts, carrying out chemical analysis of fire
debris, can compare their skills to others. A privately owned company,
Collaborative Testing Services, Inc., Herndon, VA, which is affiliated with
the Forensic Sciences Foundation, makes available standardized test
samples on an annual basis. These samples range from pure accelerants,
supplied in vials, to partially combusted matrixes which have been spiked
with accelerant mixtures. This challenging program allows each laboratory director to uncover weaknesses and take remedial action. The program has been around since 1971 and provides a variety of products to the
forensic analyst, including a section on "flammables". One of the most
useful features of this program is the final report each participant receives.
This annual report spells out technical approaches in great detail and is a
rich source of information. Some of data from the last five Round Robin
tests are summarized on the next few pages.

Most participants in these Round Robin Tests come from the U.S., but
a growing number of foreign laboratories also seem to take advantage of
this opportunity. Typically, about two thirds of all laboratories receiving
samples submit results in the form of reports. These data are compiled and
can be statistically evaluated. Because of the widely changing conditions
and the significant variation in the degree of difficulty with these samples,
it is not easy to draw general conclusions. This should perhaps be left to
each individual who chooses to take part in this type of testing. From a
technical point of view, it is rather interesting to look at current trends in
the industry. Tables 1 to 4 describe the use of analytical instrumentation

TABLE 2.

Mass Spectrometry Used in 1988–1992 Collaborative Testing Program

Instrument	1988	1989	1990	1991	1992
GC/MS alone (%)	1	5	5	12	13
GC/MS, in addition to other methods (%)	13	7	15	17	24
Number of laboratories participating	73	112	127	135	126

Note: Compiled from annual reports, Collaborative Testing Services, Herndon, VA.

TABLE 3.

Accelerant Recovery Methods, 1992 Collaborative Testing Program

Method	Percentage[a] (%)
Distillation	3
Solvent extraction	12
Headspace, cold	8
Headspace, heated	48
Adsorbent, static	50
Adsorbent, dynamic	28

Note: Compiled from 1992 annual report, Collaborative Testing Services, Herndon, VA.

[a] Response from 126 laboratories. Some laboratories used more than one method.

TABLE 4.

Column Type, 1992 Collaborative Testing Program

Column type	Percentage[a] (%)
Standard capillary	83
Wide bore capillary	19
Packed column	5

Note: Compiled from 1992 annual report, Collaborative Testing Services, Herndon, VA.

[a] Response from 126 laboratories. Some laboratories used more than one instrument.

for accelerant analysis over the last 5 years (1988 to 1992). Information is also given on the sample types provided for testing. The use of recovery methodology is only presented for the last year for which detailed data are available. The participants performed very well in the detection of common accelerants, such as gasoline, especially when the level of accelerant in the sample was high. Difficulties were encountered with accelerants which are encountered less frequently, such as light petroleum distillates, mixtures of accelerants, and turpentine. Diesel fuel and other high boiling range petroleum distillates also caused some problems. They were frequently misidentified. The latter is a reflection of the accelerant recovery technique. Most laboratories employ headspace methods, either direct or in combination with an adsorbent. Since headspace techniques have a tendency to discriminate against high boilers, it is not surprising that kerosene was often listed instead of a high boiling distillate.

It is rather surprising that laboratories which used GC/MS did not fare significantly better than institutions which only used a FID as a detector. The experience of the analyst also did not seem to make a big difference. Analysts with many years of exposure to accelerant analysis were about as likely to make mistakes as novices.[6] One of the most disturbing factors is the finding of false positives. The number of such cases is relatively small, but a few laboratories reported accelerants for sample matrixes which were not exposed to any accelerant. The reporting of a false positive is obviously a very serious error.

IV. LABORATORY PROCEDURES

A. Sample Preparation

A wide variety of methods for the analysis of residual accelerant have been developed over the years. While there is some flexibility in the selection of methods and instruments, most laboratories follow the guidelines set forth by The American Society for Testing and Materials (ASTM).[15] Current methods of chemical analysis rely on a multistep procedure in which the debris sample is first subjected to an accelerant recovery process. The sample extract is then examined by chromatography or some form of spectroscopy. In the last step, data are evaluated, usually by some technique of pattern recognition.[16] Table 5 summarizes the general process and lists some potential problems.

Sample preparation is difficult in trace analysis. Fortunately, most accelerants consist of petroleum distillates. The matrices from which volatile components are recovered are usually polymers such as carpet or wood. The analyst must choose the most suitable sample preparation

TABLE 5.

Steps in Accelerant Analysis

Step	Purpose	Procedures	Potential problems
Sample preparation	Isolation and enrichment of volatiles from matrix	Distillation, solvent extraction, headspace, static and dynamic	Volatiles may be irreversibly adsorbed, concentration may be low, partial evaporation may have taken place
Separation and/or identification	Separation and detection of accelerant-type volatiles from pyrolysis-type volatiles	Gas chromatography, GC/MS, UV, IR	Interference from matrix-type volatiles; inadequate chromatographic resolution
Data interpretation	Establishment of accelerant chromatographic profiles or spectra	Visual comparison of sample chromatogram or spectra with accelerant library; computer-based pattern recognition	Interferences from matrix-type volatiles; petroleum-based products may occur in the matrix due to pyrolysis or earlier exposure; unknown and undocumented accelerant profiles

method for the sample at hand. In the analysis of fire debris, identification is usually accomplished by pattern matching between a gas chromatographic profile of the sample and a series of accelerant standards. The particular sample preparation method chosen depends on both the type of sample and the expected accelerant. Table 6 shows an overview of the criteria for sample preparation procedures.[17] As indicated in Table 3, headspace methods are now almost universally accepted. This is quite a shift from the past where distillation and solvent extraction were considered the most reliable techniques.

Many papers have been published on the manipulation of fire-debris samples and their subsequent analysis. Procedures also reflect improvements in the state of the art of the instrumentation available. Several reviews,[18,19] and a book chapter,[20] have been presented. Some information is available that allows comparison of the relative strengths and weaknesses of the various procedures available. The selection of debris at the fire scene is obviously very critical. Unfortunately, the chemist has no input into this process. Some materials are clearly more suitable than others, from a chemist's point of view. Figure 1 represents a typical distribution of matrices submitted by fire investigators.[17] In most cases, fire debris consists of several materials, e.g., wood, padding, and carpet. Adsorptive materials, in particular carpet, are clearly preferred.

TABLE 6.

Relative Strengths and Weaknesses of Sample Preparation Procedures

Method	Strengths	Weaknesses
Direct headspace	Speed; simplicity; particularly effective for polar and highly volatile components; "clean" method	Lack of sensitivity; pronounced discrimination toward low volatility components
Adsorption/elution with solvents	High sensitivity; "clean" method	Some loss of low volatility components; polarity-based discrimination, especially toward polar substances; solvent may cover area of interest
Adsorption/elution by thermal desorption	Very high sensitivity; "clean" method	Loss of low volatility components; polarity-based discrimination, especially toward polar substances; potential artifacts; "one shot" experiment
Solvent extraction	Very effective for low volatility substances; polar and nonpolar substances can be recovered	Significant interferences from matrix; low sensitivity, unless solvent is removed; solvent purity critical; some discrimination; high demands on chromatography; same as above, but matrix interferences are reduced
Distillation	Effective recovery of polar, water soluble substances; physical isolation method; characterization possible by spectral methods (IR, UV, ...)	Very time consuming; discrimination based on volatility, solubility

The optimal sample preparation method does not only depend on the matrix of the debris, but also on the physical and chemical properties of the accelerant to be isolated. The number of potential accelerants is huge, ranging from solid oxidizers and reactive metals to conventional petroleum-based fuels. In practice, commercially available fuels, in particular gasoline, are most frequently used.[21,22] Accelerant components span from low-molecular weight alcohols to substances that are solids. The ideal sample preparation method should be capable of recovering substances over a wide boiling point range. Since accelerants may be both polar and nonpolar, it also should be effective for both. Even though quantification is usually not necessary, an indication of the approximate amount of accelerant present is desirable. Obviously, some of the requirements listed in Table 6 are incompatible. Sometimes, more than one method must be used to consolidate conflicting needs. Sample preparation may thus involve several steps.

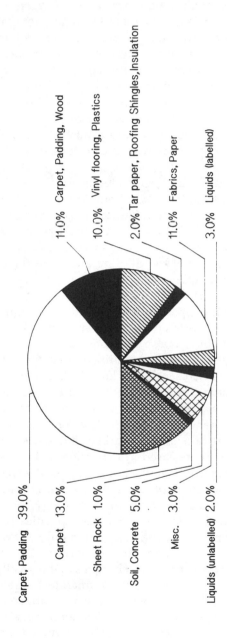

Figure 1.
Composition of matrices submitted by fire investigators for accelerant analysis. (From Bertsch, W. and Zhang, Q. W., *Anal. Chim. Acta*, 236, 183, 1990. With permission.)

The easiest and, in many ways, most straightforward method does not involve any enrichment of the sample volatiles at all. Direct headspace analysis is based on the equilibration of volatiles between the fire-debris matrix and the gas phase. It is a static method. An aliquot of the air above the sample is simply withdrawn and introduced directly into the inlet of a gas chromatograph.[23,24] Gas-tight syringes are commonly used. For a given amount of residual accelerant, the container size, amount, and temperature at sampling of matrix affect the quantity that can be detected at a given signal-to-noise ratio.[25] Direct headspace analysis is particularly useful as a screening method. It is also adequate when sufficiently large amounts of high-volatility substances are available. The most common problem with direct headspace analysis is discrimination of high boiling substances.

The major drawback of the direct-heated headspace method is a general lack of sensitivity. Adsorbent-based enrichment methods are therefore generally preferred. Adsorption/elution techniques are designed to selectively remove and trap volatiles from the headspace above a sample. The major difference between direct headspace and adsorption/elution methods is thus the removal of the diluting air. As in the direct-heated headspace method, the sample is usually heated to move volatiles from the matrix into the gas phase. The adsorbent may be inserted into the sample container or the headspace may be pulled by vacuum over a short adsorbent trap. The former is often called passive sampling.[26,27] This process is particularly simple, but it is also time consuming. Dynamic headspace methods are technically somewhat more difficult than their static counterparts, but they offer increased sensitivity and shorter sample preparation times. The choice of the adsorbent is somewhat critical. In most cases, carbon-based materials are preferred. The recovery of trapped volatiles can be effected by either solvent extraction or thermal elution. A microextraction-based method has recently been introduced.[28] The trapped volatiles are most frequently released by extraction with carbon disulfide. The adsorbent trap consists of a small amount of granular charcoal held in a modified syringe by two small plugs of glass-wool. The arrangement is an adaptation from closed-loop stripping technology, originally introduced by Grob et al. to isolate trace organic impurities from drinking water.[29] Since the volume of the charcoal bed is only a few milligrams, extraction requires only a few microliters of solvent. Down-scaling has several advantages over conventional techniques that typically use several hundred milligrams of adsorbent. With the microtechnique, solvent removal is not necessary and the entire sample can be utilized, if desired.[30] As an alternative to desorption by solvent, substances can also be released by application of heat. This method is often termed thermal elution. Synthetic porous polymers are widely used in adsorption/thermal desorption analysis. Unfortunately, synthetic polymers are prone to oxidative degradation and have inadequate capacities for low to medium mol wt range volatiles.[34] Thermal desorption from charcoal requires relatively high temperatures, a serious drawback to general headspace applications.

An interesting variation of the charcoal adsorption/thermal desorption method has been introduced. A ferromagnetic wire, coated with charcoal particles, is equilibrated over heated headspace and the adsorbed volatiles are released by heating the wire to its Curie point.[35,36] This method had been used by several laboratories in the earlier years of the collaborative testing trials, but it has not been mentioned lately. This may reflect some of the technical difficulties and the initial cost of instrumentation.

The selection of an optimum analytical method depends on many factors, of which sample preparation is only one consideration. The most important criterion for sample preparation is the success in recovering a particular accelerant type from a given matrix. Other factors such as time, level of training required, cost, and convenience must also be evaluated. One also needs to keep in mind that the heating of a sample may lead to pyrolysis. This leads to unnecessarily complex samples. Heating of samples to temperatures where pyrolysis takes place should be avoided.

B. General Instrumental Methods

After recovery of accelerant volatiles, some type of instrumental analysis must be used to characterize and identify the isolated components. In most cases, the extract is subjected to gas chromatography. Bulk methods such as UV, IR, or NMR spectroscopy are sometimes also applied.[37] The resulting data are used to determine whether a fire debris sample contains accelerant-type substances. Bulk spectral methods do not perform well with mixtures of components. IR spectroscopy is widely used for waterborne oil spills, but it cannot distinguish between substances that constitute an accelerant and those derived from pyrolysis products.[38,39] UV and fluorescence-based procedures have also been advocated. Second derivative UV spectrometry has been applied toward accelerant analysis, but it is not clear how well such a system can cope with interferences from pyrolysis.[40,41] Similar claims have been made for three-dimensional fluorescence spectroscopy.[42,43] So far, these methods have not caught on.

C. Gas Chromatography/Mass Spectrometry

Mass spectrometry and, to a lesser extent, infrared and atomic emission spectrometry can be used to identify substances. These three instruments have been successfully interfaced to gas chromatography. Mass spectrometry is particularly effective for trace level components. Advantage can be taken of common fragmentation pathways for individual substance classes which are common components in petroleum distillates. A mass spectrometer can be considered a tuneable detector for petroleum-based components. The selectivity of mass spectrometry toward a given substance class depends on the uniqueness of the fragmentation patterns. Interferences from the matrix should not produce mass fragments similar to those produced by accelerants. If several ions can be found that are characteristic for a substance class, specificity increases. Alkanes fall into this category. They produce a characteristic series of mass fragments such as m/z 43, 57, 71, 85, etc. Similar families of ions exist for other substances.

The usefulness of mass spectrometry for chromatogram simplification has been recognized early. The high cost and complexity of instrumentation has unfortunately limited wider use of this powerful technique. With the introduction of attractively priced benchtop instruments, gradual acceptance can be anticipated. Mass spectrometry has a long and successful history as an analytical tool in the petroleum industry. Unfortunately, this has not translated to the analysis of accelerants. There are relatively few publications on accelerant analysis by GC/MS prior to 1980.[44–47] The introduction of fused silica column technology facilitated the coupling of GC with MS. Several elegant applications have been presented based on quadrupole mass analyzers and fused silica column technology.[48–50] The comprehensive review by Smith is particularly noteworthy.[51] It demonstrates how mass spectrometry can discriminate against artifacts on real world samples. A book chapter by the same author provides a complete survey of the state of the art up to 1988.[52]

A mass spectrometer can be considered as a filter against interferences as well as a device that is tuneable toward target compounds. Ion profiles can be extracted from convoluted chromatograms to establish, or rule out, specific classes of substances. The term reconstructed ion chromatography describes a technique in which substance-specific information is extracted from chromatograms. Reconstructed ion chromatograms allow the analyst to focus on the target compounds. Much of the chemical noise from background components can be removed, or at least attenuated. In the case of petroleum-based substances, only a few substance classes are of interest. These are primarily alkanes and alkylbenzenes. Mass fragments have been suggested to monitor appropriate ions.[50–52] The question of selectivity toward interferences is central to the success of a mass spectrometer as an accelerant-specific detector. It is therefore necessary to examine the composition of accelerant type fluids and make comparisons to the profiles generated from matrixes. These factors are examined in the following paragraph.

D. Standards and Accelerant Classification

Petroleum-based distillates vary widely in composition and physical properties. Liquid accelerants have been categorized according to both physical principles and chemical composition. The community of forensic chemists has also taken into consideration the intended use of the material. Several schemes have been published over the years.[53-56] Table 7 is a combination of the information from several systems that has been suggested. It should be noted that a small number of petroleum-based distillates, in particular some synthetic hydrocarbons, do not fit into any of the classes.[57] The reader should also recognize that many products sold for different purposes and under different names are chemically very similar. It is thus necessary to describe the chemical composition of a distillate rather than its intended use.

Data evaluation in accelerant analysis is usually carried out by visual comparison of chromatograms. It is ironic that this relatively simple and unsophisticated process is the end product of a complex process. The analyst must be in the position to recognize an accelerant profile. This is not always easy, even with the noise-reducing capabilities of a mass spectrometer. All factors which may alter a chromatographic profile must be considered. Partial evaporation or weathering is such a variable. The distortion brought on by exposure to heat affects the human ability to recognize an accelerant. Accelerant recognition also depends on the distillate in question. Medium boiling range distillates (MPD), kerosene, and high boiling range distillates (HPD) are very similar in appearance. They display evenly spaced peaks in the chromatogram, corresponding to n-alkane homologs. Gasoline, on the other hand, does not have such easily distinguishable features. Figure 2 demonstrates that kerosene can be easily recognized even after it has lost 90% of its original components. On the other hand, 90% evaporated gasoline shows little resemblance to its original counterpart. This is shown in Figure 3. It should be noted that many of the components in gasoline, in particular alkylbenzenes, are present in significant amounts in kerosene and HPD. Normal alkanes, the characteristic compounds in kerosene and HPD, are present only in low quantities. It is thus easier to recognize kerosene in gasoline than the opposite. Some examples, illustrating these difficulties, are presented later in this chapter.

E. Matrix Contributions and Other Interferences

The recognition of petroleum-type patterns in chromatographic profiles is often hampered by the presence of interferences. Several factors must be considered in chromatogram interpretation. Chromatographic resolution, interference level, and methods of noise reduction determine how well an accelerant pattern can be distinguished. The physical presentation of chromatograms is also important. Chromatograms that consist of equally spaced peaks are more easily recognized than chromatograms

TABLE 7.

Accelerant Classification System

Class name and number	"Peak spread" based on *n*-alkane carbon numbers	Examples of products	Dominant component classes	Characteristic mass fragmental ion
1 Light distillates (LPD)	C_4–C_8	Petroleum ethers, pocket lighter fuels, solvents, Skelly solvents, VM&P naphtha	Alkanes (branched, low mol wt)	43, 57, 71, ...
2 Gasoline	C_4–C_{12}	All brands and grades of automotive gasoline including gasohol, some camping fuels	Alkanes (branched, low mol wt) Alkylbenzenes (low mol wt) Naphthalene (and low mol wt alkylderivative)	43, 57, 71, ... 91, 106, 120, ... 128, 142, 156, ...
3 Medium petroleum distillates	C_8–C_{12}	Paint thinners, mineral spirits, some charcoal starters, "dry-cleaning" solvents, some torch fuels	Alkanes (normal, interm. mol wt) Alkylbenzenes (low mol wt)	43, 57, 71, ... 91, 106, 120, ...
4 Kerosene	C_9–C_{16}	No. 1 fuel oil, Jet-A (aviation) fuel, insect sprays, some charcoal starters, some torch fuels	Alkanes (normal, medium mol wt) Alkylbenzenes (low mol wt) Naphthalene (and low mol wt alkylderivative)	43, 57, 71, ... 91, 106, 120, ... 128, 142, 120, ...
5 Heavy petroleum distillates	C_{10}–C_{12}	No. 2 fuel oil, diesel fuel	Alkanes (normal, high mol wt) Alkylbenzenes (low to medium mol wt) Naphthalenes (and similar aromatics)	43, 57, 71, ... 120, 134, 148, ... 128, 142, 156, ...
0 Unclassified	Variable	Single compounds such as alcohols, acetone, or toluene, xylenes, toluene, xylenes, isoparaffinic hydrocarbon mixtures, some lamp oils, camping fuels, lacquer thinners, duplicating fluid, others	In addition to alkanes and alkylbenzenes Alcohols Ketones Esters Terpenes	31, 45, ... 43, 58, ... 43, 73, ... 93, 136, ...

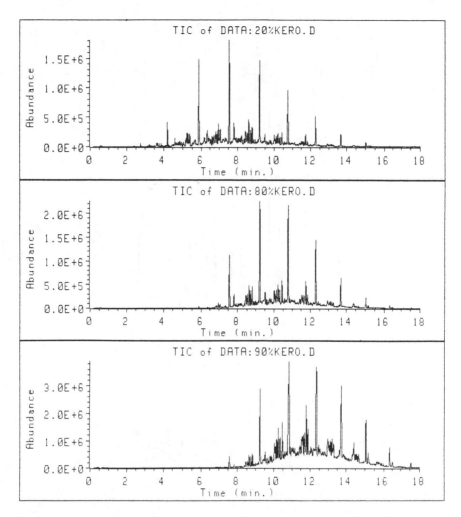

Figure 2.
Total ion chromatograms of partially evaporated kerosene. Top: 20%, middle: 80%, bottom: 90%.

consisting of randomly distributed components. Some of the pertinent factors are summarized in Table 8.

The presence of synthetic polymers in fire debris almost always causes complications. Pyrolysis can generate components which may be mistaken for petroleum-type hydrocarbons. This is especially true for carpets and similar materials. Figure 4 shows a total ion chromatogram of a sample consisting of charred carpet and carpet pad. Some of the components are listed in Table 9. The composition and distribution of a pyrolysate is strongly dependent on the conditions under which it is generated.

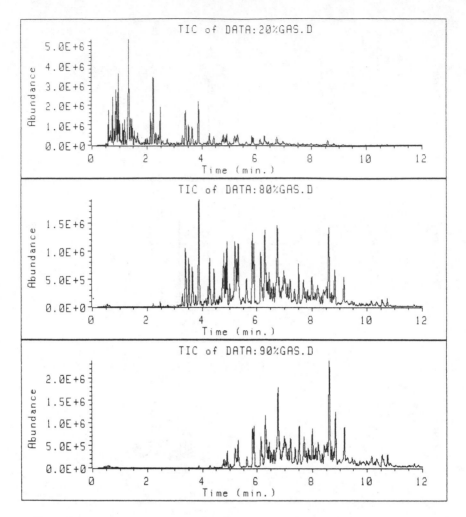

Figure 3.
Total ion chromatograms of partially evaporated gasoline. Top: 20%, middle: 80%, bottom: 90%.

Chromatographic profiles of identical materials charred under different conditions may vary widely. This is in stark contrast to the profiles of accelerants which are quite reproducible regardless of origin.[59,60] Some artifacts are generated, with some consistency, when certain matrices are exposed to heat. Figure 5 illustrates that charred carpets can display some similarities. The reader may recognize many of the major components shown in Figure 4. It is difficult to draw general conclusions about the volume and number of artifact components which are produced. Some general observations can be made, however. When styrene is generated in

TABLE 8.

Conditions Favorable for Recognition of Accelerant Profiles by Visual Comparison of Chromatograms

Factor	Condition
Chromatographic resolution	Capillary column chromatography with an efficient column
Intrinsic recognizability	Accelerants providing patterns of evenly spaced peaks
Display method	Normalized and scaled features, stacked chromatograms, side-by-side arrangement
Ratio of interferences to accelerant	Dominance of accelerant components

From Bertsch, W., Sellers, C. S., Babin, K., and Holzer, G., *HRC & CC*, 11, 815, 1988. With permission.

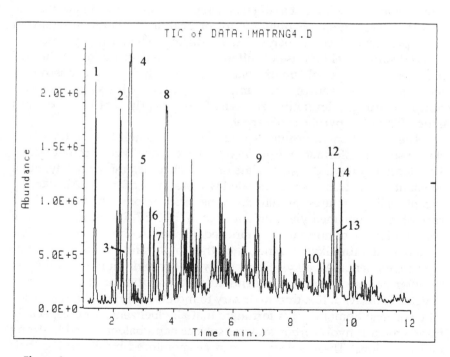

Figure 4.
Artifacts frequently encountered in charred debris from carpet/carpet pad. See Table 9 for identifications.

TABLE 9.

Artifacts Frequently Encountered in Charred Carpet/Carpet Padding

Peak no.	Compound(s)	Characteristic mass fragments (molecular ion underlined)				
1.	Toluene	$\underline{92}$	91	65	63	51
2.	Ethylbenzene	91	$\underline{106}$	51	65	77
3.	m/p-Xylenes	91	$\underline{106}$	51	77	65
4.	Styrene	$\underline{104}$	78	103	51	77
5.	Isopropylbenzene	105	$\underline{120}$	77	51	79
6.	n-Propylbenzene	91	$\underline{120}$	92	65	78
7.	1-Methyl-3-propylbenzene/ n-butylbenzene	105	$\underline{120}$	77	91	51
8.	Methylstyrene isomer	$\underline{118}$	117	103	7	51
9.	Naphthalene	$\underline{128}$	127	102	51	64
10.	2-Methylnaphthalene	$\underline{142}$	141	115	63	76
11.	1-Methylnaphthalene	$\underline{142}$	141	115	63	76
12.	$C_{15}H_{30}$-isomer	43	69	83	111	126
13.	$C_{15}H_{30}$-isomer	43	69	83	111	126
14.	$C_{15}H_{30}$-isomer	43	69	83	111	126

large amounts, the level of n-alkylbenzenes and especially of naphthalene also increases. Styrene is often the major peak in the chromatogram. In principle, artifact components can be distinguished from petroleum-derived substances because of differences in the distribution of the isomers. Isomers in petroleum distillates show a high degree of consistency regardless of their source. For example, the pattern of the xylene group is comparable in gasoline and in kerosene. This is clearly not true for xylenes generated in the pyrolysis of carpet.

The major benefit of mass spectrometry is its ability to extract substance-specific information from an otherwise convoluted chromatogram. The selectivity toward a given class of substances is, unfortunately, not a constant factor. This can be illustrated with a simple example. All alkanes, except cyclic structures, present the same series of ions which differ from each other by one methylene unit. Ions of m/z 43, 57, 71, etc. are therefore good indicators of normal and branched alkanes. A similar series of ions does not exist for alkylbenzenes, the other important class of accelerant-type components. All alkylbenzenes have the tropylium ion in common, but its response is strongly dependent on the particular structure from which it is derived. It is thus necessary to monitor a different ion for each alkylbenzene homolog. It is fortunate that alkylbenzenes, and especially fused aromatic hydrocarbon systems such as naphthalene, provide strong molecular ions. These substances can be monitored by observing their molecular ions. The major problem in analysis is the complexity and volume of the data sets. A significant effort is required for data treatment.

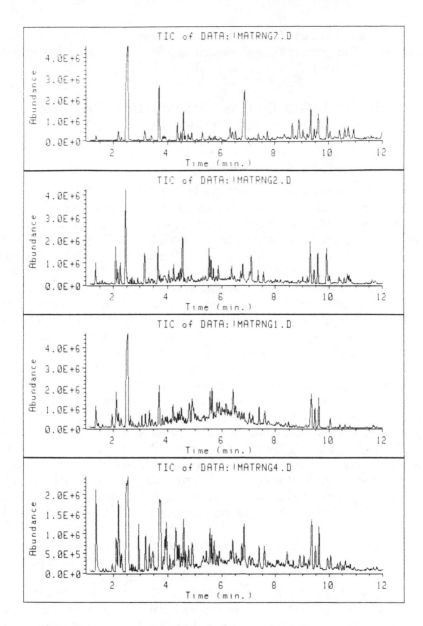

Figure 5.
Comparison of profiles from several charred carpets/carpet pads.

It is fortunate that this difficulty can be overcome with automated data processing. This process is briefly discussed in the following paragraph.

V. THE PRACTICE OF GAS CHROMATOGRAPHY/MASS SPECTROMETRY

A. General Principles of Data Reduction

The volume of data generated in a typical GC/MS run is very large, and extraction of diagnostic ion profiles can be quite time consuming. An expert-based system has recently been described which automatically extracts the diagnostic components of a sample and displays them with the profiles of standards, side by side.[58-62] Figure 6 shows a total ion chromatogram of a simulated arson sample and Figure 7 shows a diagnostic report for the same sample. The figure is very complex and contains a great deal of information. The analyst is provided with diagnostic information which confirms or rules out the presence of various components. Some of the boxes appear too small to be useful for interpretation in this reproduction. When displayed full size, all features become meaningful. The reader is reminded that it is not always necessary to discern a particular pattern. It is sometimes only necessary to confirm the presence of a single peak in a given retention time window. If a substance is not present, a given accelerant can be ruled out. The particular arrangement and the rationale of presenting diagnostic information in this manner has been described in detail.[6,59,63] This kind of presentation requires custom software which has to be written by the user and is dependent on the configuration of the instrument used. MACRO programming is widely used to carry out repetitive tasks and speed up time-consuming steps in data reduction. It can readily be implemented on any modern GC/MS instrument.

The display of diagnostic ions for the sample whether done manually or in an automated manner is the first step in the decision making process. The volume of data of potential interest is very large. The analyst has to decide what kind and how much data is needed. The time required for data processing and printout needs to be balanced against the desire to present the data as completely as possible. A one-page summary is perhaps a reasonable compromise. It may be beneficial to briefly summarize the factors that are important in generating a meaningful report.

The requirements for data evaluation depend on whether interpretation is carried out by visual inspection or by a computerized pattern recognition program. It is always advisable to have available a visual representation even if final data reduction is carried out by a sophisticated

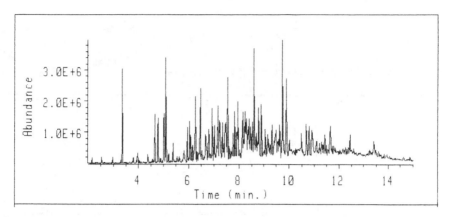

Figure 6.
Total ion chromatogram of charred carpet/carpet pad, spiked with 50% evaporated gasoline.

pattern recognition program. To fulfill human needs it is important to provide side-by-side plots of diagnostic ions. This is particularly helpful if patterns are complex and irregular, such as with higher alkylbenzene homologs. Table 10 discusses some of the factors that are important in designing the format for a visual diagnostic report. The particular layout of the information is not important. It depends on personal preference and on the equipment available. Some diagnostic ions such as the C_2-alkylbenzenes, naphthalene, and methylnaphthalenes represent very simple patterns which can easily be memorized by the observer. Others are more complex and side-by-side comparison with appropriate standards is not just a matter of convenience but increases the level of confidence. Inclusion of information on potential artifacts is particularly important. Selected ion chromatograms characteristic for alkenes and styrenes are helpful in determining the level of interferences that must be dealt with. If styrenes are present at a high concentration, the levels of toluene, n-alkylbenzenes, naphthalenes, and 2-methylnaphthalene are likely to be increased. Figure 8 shows a chromatogram of a charred sample which does not contain an accelerant. The chromatogram corresponds to the matrix used for the simulated arson sample shown in Figure 6. Because of the complexity of the chromatogram it is very difficult to detect the presence of an accelerant, especially one which is not distinguished by easily recognizable features. The differences between the simulated arson sample shown (Figure 6) and the matrix (Figure 8) are very small. When the same matrix was spiked with kerosene, accelerant recognizability improved significantly. Figure 9 shows a comparison of some of the ion chromatograms displayed in the diagnostic reports of the samples shown in Figures 6 and 8. The figure has been assembled for demonstration purposes by cutting, pasting, and enlarging some of the diagnostic ions.

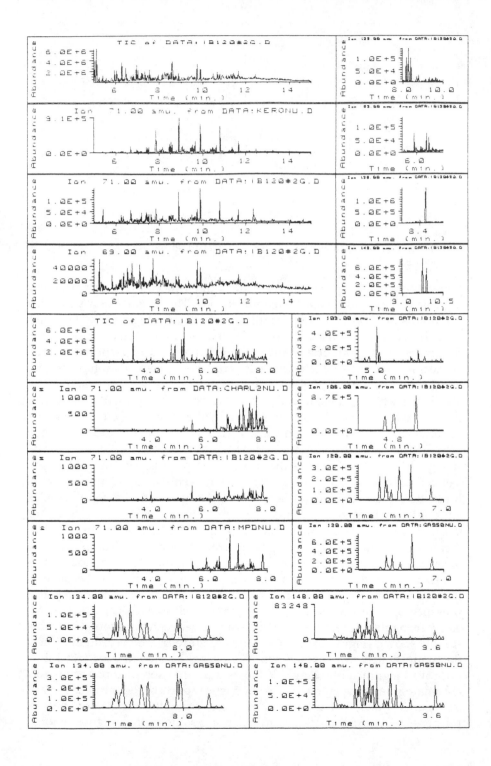

Figure 7.
Diagnostic report for spiked fire debris sample shown in Figure 6.

TABLE 10.

Design Considerations for a Comprehensive Diagnostic Report

Feature displayed	Rationale
Total ion chromatogram for entire sample; expansion of the early part of the sample chromatogram	A broad overview of the general pattern; provide detail of the narrow and crowded section corresponding to highly volatile components
Extracted ion profiles for C_2–C_5 alkylbenzenes of the samples	Diagnostic ions for most distillates, in particular for gasoline
Extracted ion profiles for C_3–C_5 alkylbenzenes of a gasoline standard	Side-by-side comparison of complex irregular patterns to facilitate visual pattern recognition
Extracted on profile m/z 71 of the sample	Diagnostic ion for normal and branched alkanes; diagnostic indicators for LPD, kerosene, and IIPD
Extracted ion profile m/z 71 of a kerosene standard	Side-by-side comparison for verification of minor branched alkanes
Expansion of extracted ion profile m/z 71 for the early part of the sample	Provide detail for LPD-type alkanes
Expansion of extracted ion profile m/z 71 of a naphthenic distillate	Side-by-side comparison of complex irregular patterns to facilitate visual pattern recognition
Extracted in profile of m/z 69 of the sample	Provide an indication of the level of interferences
Extracted ion profile of m/z 103 of the sample	Provide information on styrenes
Extracted ion profiles of m/z 93 of the sample	Provide information on terpenes

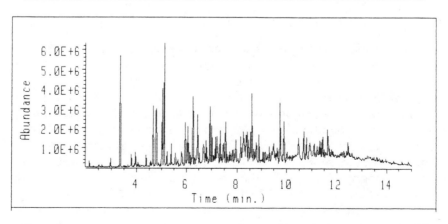

Figure 8.
Total ion chromatogram of charred carpet/carpet padding with no accelerant. The matrix is the same as in Figure 6.

m/z 120   m/z 134   m/z 142

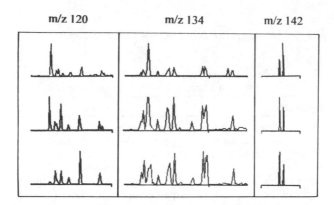

Figure 9.
Selected ion chromatograms of diagnostic components for gasoline. Top: matrix of
charred carpet/carpet pad from Figure 8, middle: spiked matrix from Figure 6,
bottom: 50% evaporated gasoline, m/z 120 = C_3 benzenes, m/z 134 = C_4 benzenes,
m/z 140 = C_5 benzenes, m/z 103 = styrenes, m/z 142 = methylnaphthalenes.

Slight variations are visible between the ion profiles displayed by the
spiked sample and the gasoline standard. These are relatively small and
reflect a small contribution from the matrix for the spiked sample. The
situation is different for the charred matrix. It can be seen that small
amounts of diagnostic ions are produced in the process of pyrolysis, but
the profiles are dominated by only a few ions. Distortion of the ion profiles
is quite obvious. Additional information can be extracted by comparing
the magnitude of the ion responses in each box. Inspection of the scale
printout on each y-axis provides response ratios which aid in determining
artifacts.

B.   Aspects of Automation

The previous section provided information on interpretation of simulated fire debris samples. The interpretations were done by visual comparison of the total ion profile and several extracted ion profiles of the
sample with those of accelerant standards. This process is time consuming
and manual. Few attempts have been made to incorporate computer-
based pattern recognition. A GC/MS expert system has been described.[6,58]
Figure 10 presents a flow diagram. The data acquired from the sample are
compared with those stored in the computer for various accelerants. Based
on chromatographic as well as mass spectra criteria, the system tests
various possibilities. Figure 11 shows a diagram of the principal steps
necessary to establish the presence of gasoline. The final product of the
program is a table that provides match factors. The system is instructed to

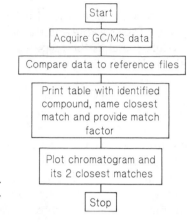

Figure 10.
Flow diagram of a GC/MS-based expert system. (From Holzer, G. and Bertsch, W., *Anal. Chim. Acta*, 259, 225, 1992. With permission.)

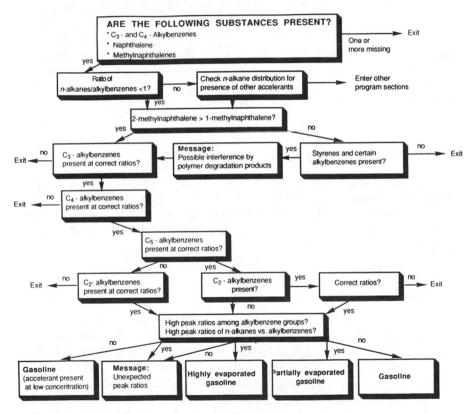

Figure 11.
Flow diagram of the principal steps in the automated evaluation by the expert system. The identification of gasoline is demonstrated. (From Holzer, G. and Bertsch, W., *Anal. Chim. Acta*, 259, 225, 1992. With permission.)

plot, side by side, a total ion chromatogram or a selected ion chromatogram of the sample and the accelerant standard from the library that most closely matches the responses from the sample under consideration. The algorithm for the program is straightforward, but it is not easily implemented because of its complexity. The program can also be instructed to plot mass spectra of diagnostic components in the sample and standard. Figure 12 displays an example for a rather unusual accelerant.

It is interesting to test the limits of an expert system with samples containing variable amounts of accelerant in the presence of a constant background. The experimenter can set up conditions where the expert system will ultimately fail. One can imagine how the selected ion profiles in Figure 9 change as the level of gasoline decreases. It is very difficult to define conditions, such as a range of peak ratios among isomers, which will always discriminate against artifacts or always detect the presence of low concentrations of an accelerant. The expert system is very useful in sorting out cases that are strongly positive or strongly negative for accelerant. In situations where matrix contributions cause significant distortions, the analyst must come to the rescue. A manual inspection of the entire data set may be needed in critical cases. The final decision must always rest with the forensic chemist.

VI. SELECTED EXAMPLES

Accelerants encountered in practice fall into only a few categories. Some of these are more easily detected than others. The following examples provide guidelines for interpretation of general response chromatograms and selected ion chromatograms for the major classes of accelerants.

A. Gasoline

Gasoline consists of well above 250 components above the 10 ppm level. It is probably one of the most complex accelerants. Modern gasolines are enriched in aromatic hydrocarbons, whose presence is instrumental in the identification of the accelerant in fire debris samples. The distribution of the alkylated aromatic compounds can be used to determine the degree of evaporation or weathering the sample has undergone. The most important diagnostic compounds for positive identification of gasoline in fire debris samples are C_2- to C_5-alkylbenzenes, naphthalene, and methylnaphthalenes. Aliphatic hydrocarbons are also important, especially in fresh gasolines. The total ion current (TIC) of a gasoline standard, an 80% evaporated gasoline standard, and fire debris sample 1 are shown in

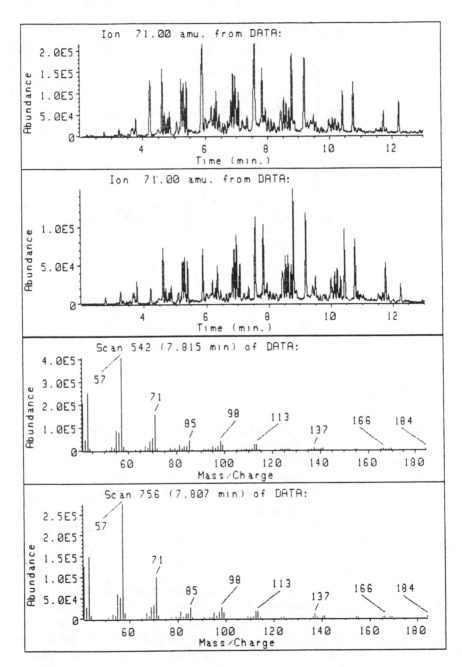

Figure 12.
Ion chromatograms in m/z 71 and mass spectra from a liquid found in a storage
shed at a fire scene, and a debris sample. (From Bertsch, W., Zhang, Q. W., and
Holzer, G., *HRC*, 13, 597, 1990. With permission.)

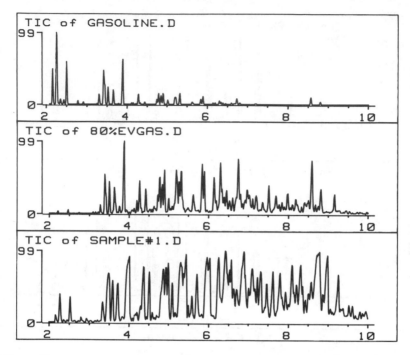

Figure 13.
Total ion chromatogram of a gasoline standard, 80% evaporated gasoline, and fire
debris sample 1.

Figure 13. Visual examination of the chromatograms of sample 1 reveals
a depletion of the low boiling C_2-alkylbenzenes and a relative increase of
higher boiling components. This is most likely caused by weathering of
exposure to heat. A comparison with the chromatographic profile of an
80% evaporated gasoline standard shows good agreement between the
alkylbenzene pattern of the sample and standard.

The distribution of the various groups of aromatic hydrocarbons is
best visualized by single ion current (SIC) profiles using the molecular
ions of m/z 106, 120, 134, and 148 for alkylbenzenes, m/z 128 for naphthalene, and 142 for the methylnaphthalenes (Figure 14). The C_2-alkylbenzenes
are separated into ethylbenzene, co-eluting p- and m-xylene, and o-xylene.
The ethylbenzene peak may be enhanced since it is a thermal decomposition product of some plastic materials. The C_3-alkylbenzenes consist of
eight isomers which have a very characteristic distribution. The pattern is
important for the identification of partially evaporated accelerants, where
the C_2-alkylbenzenes are missing. The propylbenzene isomers (peak 5) are
sometimes enhanced in fire debris samples with thermally decomposed
polymers. As the alkyl chain increases, the chromatographic profiles become more complex. Among the 21 possible isomers of C_4-alkylbenzenes,

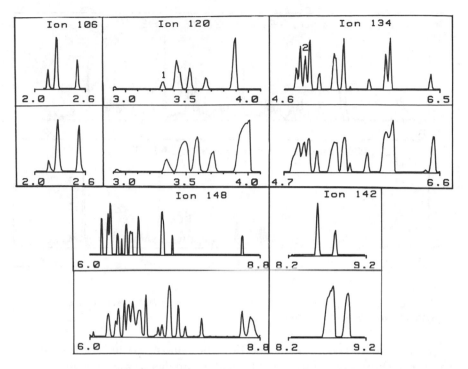

Figure 14.
Comparison of single ion chromatograms of gasoline (top chromatograms) and fire debris sample 1 (bottom chromatograms). Ion 106: C_2-alkylbenzenes; ion 120: C_3-alkylbenzenes; ion 134: C_4-alkylbenzenes; ion 148: C_5-alkylbenzenes; ion 142: methyl-naphthalene isomers.

approximately 11 compounds are sufficiently abundant to be useful for the identification of gasoline. In the presence of thermal decomposition products from polymers and plastics, peak 14 is enhanced. The last group of diagnostic alkylbenzenes, the C_5-alkylbenzenes, is usually well defined in samples which have undergone extensive evaporation. Only minor distortions of the pattern are usually observed in the presence of outgasing substances from burned plastics or polymers. The two methylnaphthalene isomers are very critical for the positive identification of gasoline. Acceptable ratios of 1-methyl-naphthalene/2-methylnaphthalene must be smaller than 1. Naphthalene by itself is not a specific indicator for gasoline; however, in its absence, gasoline can be ruled out as an accelerant.

Additional evidence for the presence of gasoline is available from the alkane distribution. The profile is difficult to detect in flame ionization detector chromatograms (FID) but can easily be visualized in a m/z 71 SIC chromatogram. The m/z 71 mass spectral fragment is diagnostic for n-alkanes and higher branched alkanes. The dominating aromatic compounds filtered out and the characteristic unimodal distribution of n-alkanes

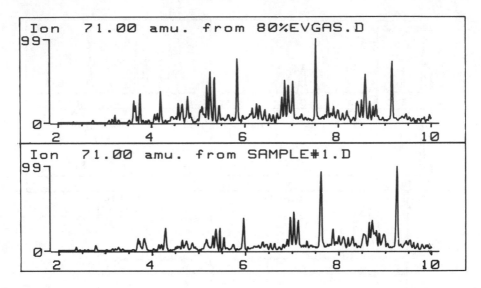

Figure 15.
Alkane specific single ion chromatogram (m/z 71) of an 80% evaporated gasoline standard and fire debris sample 1.

becomes visible. The *n*-alkane and branched alkane pattern of fire debris sample 1, shown in Figure 15, is typical for an evaporated gasoline standard. The alkane pattern is less pronounced in unevaporated samples. The degree of accelerant evaporation can be determined from peak ratios of alkylbenzene isomers from different groups and *n*-alkane/alkylbenzene ratios.[6]

B. Medium Petroleum Distillates (MPD)

MPDs comprise an accelerant class which includes charcoal lighter fluids, mineral spirits, and paint thinners. Despite the variety of commercial products, the chemical composition of MPD from different refineries is quite similar. The principal constituents are *n*-alkanes in the range of $n\text{-}C_8$ to $n\text{-}C_{12}$. The distribution is unimodal with $n\text{-}C_{10}$ or $n\text{-}C_{11}$ as major components. The chromatographic pattern of a MPD standard is compared to the elution profile of fire debris sample 2 (Figure 16). For positive identification, at least three consecutive *n*-alkanes between $n\text{-}C_9$ and $n\text{-}C_{12}$ should be present. Additional proof for the presence of MPD in sample 2 is provided by branched alkanes, which elute between consecutive pairs of *n*-alkanes. The chromatographic profile of fire debris sample 3 points also toward the presence of MPD. However. the depletion of the low boiling components and a shift of the pattern toward higher boiling com-

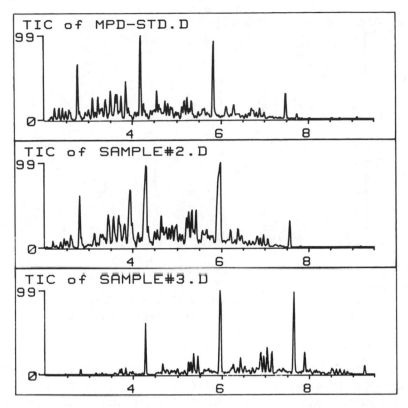

Figure 16.
Total ion chromatogram of a medium petroleum distillation product (MPD-STD)
and fire debris samples 2 and 3.

pounds indicate strong evaporation of the accelerant, possibly because of
exposure to heat or prolonged weathering. Other alternations of the
n-alkane pattern may be caused by thermal decomposition products of
plastic materials. In such cases, higher n-alkanes (n-C_{12} or n-C_{15}) are added
to the MPD pattern, which then may resemble a kerosene profile. The
interference can be identified because of the absence of clearly defined
clusters of branched alkanes between the higher n-alkanes.

In addition to charcoal starters, mineral spirits, and paint thinners,
MPDs are used as solvents for certain insect sprays and may be present in
newly manufactured materials such as tile cements, glues, and protective
barriers. Other petroleum-based products which have boiling ranges similar to MPD are naphthenic charcoal lighter fluids. Lacking n-alkanes, the
major components are naphthenic hydrocarbons. Figure 17 compares the
chromatographic profiles of a naphthenic and a MPD-based charcoal
lighter fluid.

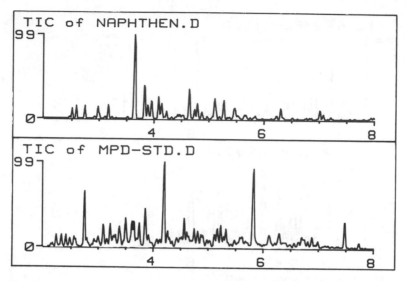

Figure 17.
Total ion chromatogram of a naphthenic charcoal lighter fluid standard
(NAPHTHEN) and a medium petroleum distillation product (MPD-STD).

C. Kerosene

The partial chromatograms of a kerosene standard and fire debris
samples 4 and 5 are shown in Figure 18. One of the characteristic features
of the hydrocarbon composition of kerosene is the unimodal distribution
of n-alkanes in the range of n-C$_8$ to n-C$_{17}$ and the clusters of branched
alkanes which elute between consecutive pairs of n-alkanes. The chromatographic profile of sample 4 exhibits the distinct hydrocarbon pattern
of kerosene. The depletion of the low boiling components eluting at the
beginning of the chromatogram indicates partial evaporation of the
accelerant. The chromatogram of sample 5, which was obtained at the
same fire scene as sample 4, resembles the kerosene standard, except for
the intense olefin peaks, which elute before each n-alkane. These compounds are most likely thermal decomposition products of household
organic polymers. When polyethylene or polypropylene is heated to high
temperatures, a series of n-alkanes, mono- and di-unsaturated olefines are
formed; however, very little or no branched alkane materials are released.
The presence or absence of kerosene in sample 5 can be determined from
a single ion plot of m/z 71 which records only saturated alkanes. The
bottom chromatogram in Figure 18 shows what might be construed as an
underlying kerosene profile. In most cases, however, it is not possible to
identify kerosene in the presence from outgasing products from decomposed polymers.

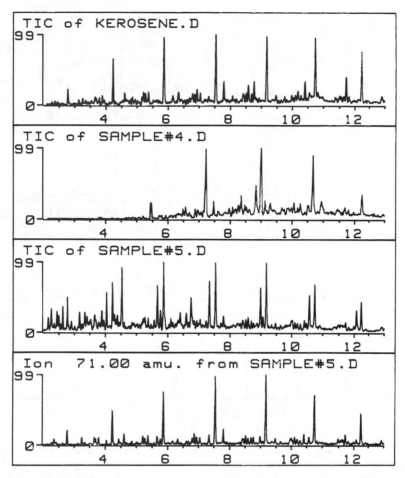

Figure 18.
Total ion chromatogram of a kerosene standard and fire debris samples 4 and 5. The bottom chromatogram shows an alkane-specific single ion chromatogram (m/z 71) of sample 5.

D. Heavy Petroleum Distillates (HPD)

Heavy petroleum distillates such as diesel fuel and no. 2 heating oils have a very broad range of n-alkanes, extending from n-C_{10} to n-C_{25}. Similar to kerosene, there are clusters of branched alkanes eluting between the n-alkanes. Depending on the nature of the fire debris and the sample preparation method, recovery of the higher boiling hydrocarbons is often incomplete. In such cases the pattern can easily be mistaken for kerosene. For positive identification of HPD, five consecutive n-alkanes have to be present in addition to the two isoprenoid hydrocarbons, pristane and

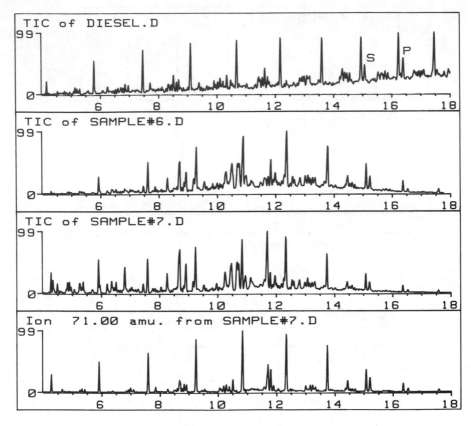

Figure 19.
Total ion chromatogram of diesel fuel standard and fire debris samples 6 and 7. S: pristane, P: phytane. The bottom chromatogram shows an alkane-specific single ion chromatogram (m/z 71) of sample 7.

phytane. These two compounds are natural constituents in crude oil and are generally not present in kerosene. Similar to other petroleum distillation products, the criteria for positive identification of HPD also includes branched alkanes. Such compounds are, for example, not present in waxes, which may have a *n*-alkane distribution that resembles a severely evaporated HPD.

Figure 19 shows the hydrocarbon profile of a HPD standard (diesel oil) and fire debris samples 6 and 7. The samples were prepared by the heated headspace method, which accounts for the poor recovery of the high boiling components. Despite these shortcomings, the broad range of *n*-alkanes and the presence of pristane and phytane clearly identify HPD as an accelerant in sample 6. There is no distortion in the pattern of the diesel oil since the standard was injected directly as a liquid onto the column. The advantage of single ion profiles in the identification of petro-

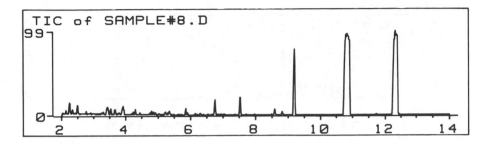

Figure 20.
Total ion chromatograms of a lamp oil containing fire debris sample 8. The two major peaks are n-C_{14}- and n-C_{15}-alkanes.

leum distillate-based accelerants is demonstrated in the bottom chromatogram of Figure 19. The presence of HPD in sample 7 is not obvious, based on the pattern of the TIC which is similar to a FID chromatogram. The profile is, however, clearly visible in the m/z 71 plot, which filters out all aromatic and unsaturated hydrocarbons.

E. Lamp Oils

The hydrocarbon composition of lamp oils consists of n-alkanes in the range of nC_{13} to nC_{18}. The pattern is usually dominated by two consecutive n-alkanes. Figure 20 shows the volatiles released from fire debris sample 8. The major components are n-C_{14} and n-C_{15} alkanes which originated from the lamp oil present in the sample. Some plastic materials have thermal outgasing products which can be mistaken for lamp oil. For example, in the combustion of vinyl strips, a series of n-alkanes in the range of n-C_{13} to n-C_{15} may be generated. The presence of such plastic decomposition products can be recognized if other compounds such as chlorinated substances are detected.

F. Turpentine

Turpentine is a by-product in the pulp and paper industry and consists mostly of terpene hydrocarbons. Since it is derived from wood, terpenes are commonly detected in fire debris samples which contain woody materials. Commercial turpentines may differ considerably in their chemical composition. The top chromatogram in Figure 21 shows the TIC of a turpentine standard. The chromatogram displayed below is a single ion plot of m/z 93, which is a specific mass spectral fragment common to most terpenes present in turpentine. The single ion plot identifies only four major terpenes in the formulation. The remaining substances are nonterpenoid additives. The chromatographic profiles (TIC and terpene

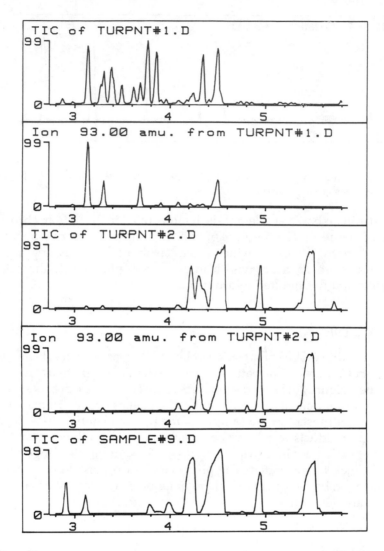

Figure 21.
Total ion and single ion chromatograms of two turpentine standards (TURPNT#1 and TURPNT#2). The bottom chromatogram shows a terpene-specific single ion chromatogram (m/z 93) of sample 9.

specific m/z 93 plot) of a second turpentine standard shown in Figure 21 suggest that no additives have been added to the formulation. The volatiles of fire debris sample 9 are similar to the profile of the second standard. Since the sample contained no woody materials, the terpenes must have originated from turpentine.

VII. FUTURE DEVELOPMENTS

Progress in science is often a reflection of the tools available for exploration. Organic trace analysis and, in particular, the detection of accelerants in fire debris, are no exception. The technology to build a dedicated accelerant detector is already available. Portable mass spectrometers and sophisticated vapor sampling devices are increasingly employed in field work especially for the evaluation of toxic emissions at Superfund sites.[65] The key is miniaturization of instrumentation. A portable mass spectrometer weighing a mere 38 kg has been described.[66] Membrane introduction mass spectrometry is an attractive approach to the analysis of sample vapors. It has good sensitivity, but it does not seem to have adequate specificity to distinguish between substances from accelerants and matrices.[67] An instrument of this type functions by continuously monitoring the atmosphere. No chromatographic separation steps are included. The situation can be improved by addition of a very short capillary column, in the order of 1 m. A system capable of sampling rates at a turnaround of a few seconds has been described for the analysis of ppm levels of atmospheric vapors.[68] A solution to improve specificity is the addition of ion/molecule reactions, following initial ionization in a mass spectral ion source. This technique commonly named MS/MS is both fast and has the good potential for high specificity. At the present time, MS/MS is not sufficiently developed for routine analysis in cases where matrix contributions change drastically. It is not unreasonable to project that a MS/MS-based portable device for accelerant analysis will become available in the future.

ACKNOWLEDGMENTS

We would like to thank Evelyn Jackson for typing this manuscript on very short notice.

REFERENCES

1. Phillips, C. C. and McFadden, D. A., *Investigating the Fire Ground*, Dun & Bradstreet, New York, 1982.
2. O'Connor, J. J., *Practical Fire and Arson Investigation*, Elsevier, New York, 1986.
3. DeHaan, J. D., *Kirk's Fire Investigation*, 3rd ed., John Wiley & Sons, New York, 1983.
4. Carroll, J. R., *Physical and Technical Aspects of Fire and Arson Investigation*, Charles C Thomas, Springfield, IL, 1979.

5. Roblett, C. L., McKechnie, A. H., and Lundy, V., *Investigation of Fires*, Brady Publ., Englewood Cliffs, NJ, 1988.
6. Bertsch, W., Holzer, G., and Sellers, C. S., *Chemical Analysis for the Arson Investigator and Attorney*, Hüthig, Heidelberg, 1993.
7. U.S. Department of Justice, *Law Enforcement Assistance Administration, Survey and Assessment of Arson and Arson Investigation*, Aerospace Corporation, Law Enforcement Development Group, Pub. No. ATR76 (791805) 2, (October, 1976), pp. 79–87.
8. Microsensor Technology, Inc., Fremont, CA.
9. Tontarski, R. E., *J. Forensic Sci.*, 2, 440, 1983.
10. Kinard, W. D. and Midkiff, C. R., *J. Forensic Sci.*, 36, 1714, 1991.
11. Dietz, W. R. and Mann, D. C., *Fire Arson Investigator*, 39, 26, 1988.
12. Camp, M. J., *On Being an Expert in Forensic Science*, Davis, G., Ed., American Chemical Society, Washington, D.C., 1986.
13. Cantor, B. J., Tips for the expert witness, in *The Expert and the Law*, Gorman, M. J., Ed., Lawrenceville, NJ, 1986, 5.
14. Kantrowitz, S. B., *J. Forensic Sci.*, 26, 1414, 1981.
15. Standard Test Method For Flammable Or Combustible Liquid Residues In Extracts From Samples Of Fire Debris By Gas Chromatography, ASTM Method E 1387–90, ASTM, 1916 Race St., Philadelphia, PA.
16. Bertsch, W. and Holzer, G., *Eur. Chromatogr. News*, 2, 3, 12, 1988.
17. Bertsch, W. and Zhang, Q. W., *Anal. Chim. Acta*, 236, 183, 1990.
18. Lee, H. C., Desio, P. J., and Gaensslen, R. E., *Forensic Science*, 2nd ed., Davies, G., Ed., American Chemical Society, Washington, D.C., 1986, 279.
19. Caddy, B., Smith, F. P., and Macy, J., *Forensic Sci. Rev.*, 3, 57, 1991.
20. Midkiff, C. R., in *Arson and Explosive Investigation in Forensic Science Handbook*, Saferstein, R., Ed., Prentice Hall, Englewood Cliffs, NJ, 1982, 222.
21. Kuntz, R. L., *Arson Anal. Newsl.*, 2, 6, 1978.
22. DeHaan, J. D., *Arson Anal. Newsl.*, 2, 39, 1978.
23. Camp, M., *Anal. Chem.*, 52, 422A, 1980.
24. Poll, K. D. and Keller, E., in *Applied Headspace Gas Chromatography*, Kolb, B., Ed., Heyden, London, 1980, 165.
25. Kurz, M. E., Jakacki, J., and McCaskey, B., *Arson Anal. Newsl.*, 8, 1, 1984.
26. Kobus, H. J., Kirkbride, K. P., and Maehly, A. J., *Forensic Soc.*, 25, 307, 1987.
27. Dietz, W. R., ATF, San Francisco Center, personal communication.
28. Sellers, C. S., Ph.D. dissertation, University of Alabama, 1989.
29. Grob, K. and Zurcher, F., *J. Chromatogr.*, 117, 285, 1976.
30. Grob, K., *Classical Split and Splitless Injection in Capillary GC*, Hüthig, Heidelberg, 1988.
31. Brown, R. H. and Purnell, C. J., *J. Chromatogr.*, 178, 79, 1979.
32. Holzer, G., Shanfield, H., Zlatkis, A., Bertsch, W., Juarez, D., Mayfield, H., and Liebich, H. M., *J. Chromatogr.*, 142, 755, 1977.
33. Jennings, W. G. and Filsoof, M. J., *Agric. Food Chem.*, 25, 440, 1977.
34. Andrasco, J., *Forensic Sci.*, 28, 330, 1983.
35. Baldwin, R. E., *Arson Anal. Newsl.*, 1, 9, 1977.
36. Twibell, J. D., Horne, J. M., and Smalldon, K. W., *Chromatographia*, 14, 366, 1981.
37. Stone, T. C., Lamonte, J. V., Fletcher, L. A., and Lowry, W. T., *J. Forensic Sci.*, 23, 78, 1978.
38. Brown, C. W., Lynch, P. F., and Ahmadjian, M., *Appl. Spectrosc. Rev.*, 9, 223, 1975.
39. Thaman, R. N., *Arson Anal. Newsl.*, 3, 9, 1979.
40. Meal, L., *Anal. Chem.*, 58, 834, 1986.
41. Meal, L., *Fire Arson Investigator*, 39, 1, 5, 1988.
42. Fulton, E. A. and Meloan, C. E., *J. Forensic Sci.*, 31, 1086, 1986.

43. Alexander, J., Mashak, G., Kapitam, N., and Siegel, J. A., *J. Forensic Sci.*, 32, 72, 1987.
44. Zoro, J. A. and Hadley, K. J., *Forensic Sci. Soc.*, 16, 103, 1976.
45. Smith, R. M., *Am. Lab.*, 10, 53, 1978.
46. Mach, M. A., *J. Forensic Sci.*, 22, 348, 1977.
47. Stone, I. C., Lomonte, J. N., Fletcher, L. A., and Lowry, W. T., *J. Forensic Sci.*, 23, 78, 1978.
48. Smith, R. M., *J. Forensic Sci.*, 28, 318, 1983.
49. Trimpe, M. A. and Tye, R., *Arson Anal. Newsl.*, 7, 2, 26, 1983.
50. Kelly, R. L. and Martz, R. M., *J. Forensic Sci.*, 29, 714, 1984.
51. Smith, R. M., *Anal. Chem.*, 4, 1399A, 1982.
52. Smith, R. M., *Arson Analysis by Mass Spectrometry in Forensic Mass Spectrometry*, Yinon, J., Ed., CRC Press, Boca Raton, FL, 1989.
53. Nowicki, J., *J. Forensic Sci.*, 35, 1064, 1990.
54. Analysis Classification Scheme, *Arson Anal. Newsl.*, 6, 57, 1982.
55. Guidelines for laboratories performing chemical and instrumental analysis of fire debris samples, *Fire Arson Investigator*, 38, 4, 45, 1988.
56. Bertsch, W. and Sellers, C. S., *HRC & CC*, 9, 657, 1986.
57. Gartun, M. A., *Arson Anal. Newsl.*, 6, 65, 1982.
58. Bertsch, W., Sellers, C. S., Babin, K., and Holzer, G., *HRC & CC*, 11, 815, 1988.
59. Holzer, G. and Bertsch, W., *Anal. Chim. Acta*, 259, 225, 1992.
60. Bertsch, W., Sellers, C. S., and Holzer, G., *LC-GC*, 6, 1000, 1988.
61. Wineman, P. L. and Keto, R. O., *Anal. Chem.*, 63, 1964, 1991.
62. Wineman, P. L., ATF National Laboratory Center, in *Symp. Recent Advances in Arson Analysis and Detection*, Las Vegas, NV, 1985.
63. Bertsch, W., Wells, C., and Holzer, G., *Proceedings of the 10th International Symposium on Capillary Chromatography*, Sandra, R., Ed., Hüthig, 1989, 1302.
64. Bertsch, W., Zhang, Q. W., and Holzer, G., *HRC*, 13, 597, 1990.
65. Trainor, T. M. and Lankinen, F. H., Paper 144, Proc. 26th Pittsburgh Conference, Atlanta, GA, 1989.
66. Newsman, A. R., *Anal. Chem.*, 63, 641A, 1991.
67. Kotiaho, T., Lauritsen, F. R., Choudhurg, T. K., Cooks, R. G., and Tsao, T. G., *Anal. Chem.*, 63, 875A, 1991.
68. Arnold, N. S., McLennen, W. H., and Meuzelaar, H. L. C., *Anal. Chem.*, 63, 299, 1991.

Chapter 5

FORENSIC APPLICATIONS OF PYROLYSIS-MASS SPECTROMETRY

Thomas O. Munson

CONTENTS

0-8493-8252-1/95/$0.00+$.50
© 1995 by CRC Press Inc.

I. INTRODUCTION

This chapter attempts to cite, discuss, and correlate all published forensic applications of pyrolysis-mass spectrometry (Py-MS), including pyrolysis-gas chromatography/mass spectrometry (Py-GC/MS). In this context, the term "forensic applications" is taken to mean those instances where these analytical techniques were used for the examination of materials which were evidence in a court of law or where a published report had the purpose of demonstrating how the analytical technique might be used with such evidential materials.

The technical aspects of Py-MS and Py-GC/MS will be discussed in detail later in this chapter, but, at this point, brief descriptions of the techniques might be helpful. In this context, the term pyrolysis refers to the cleavage of large, nonvolatile polymeric molecules into smaller, volatile molecules by the rapid input of thermal energy. The resultant mixture of volatile molecules can be analyzed directly using mass spectrometry (Py-MS), or the mixture can be passed through a gas chromatograph to effect some separations before being analyzed by mass spectrometry (Py-GC/MS). Both types of analytical pyrolysis have particular strengths and weaknesses when applied to materials of interest to the forensic scientist.

One reviewer of forensic analytical pyrolysis observed that, while analytical pyrolysis was a technique used in most forensic science laboratories, the published literature was relatively limited because few took the time to publish their experiences.[1] The published literature describing forensic Py-MS and Py-GC/MS is especially sparse because these techniques comprise but a small slice of the whole of forensic analytical pyrolysis, which is dominated by pyrolysis-gas chromatography (Py-GC). The literature search for this review of forensic pyrolysis mass spectrometry (which includes the first few months of 1993) located about 35 publications from forensic science laboratories spanning the time period 1969 to 1993.

II. HISTORY OF FORENSIC PYROLYSIS-MASS SPECTROMETRY

Several authors[2,3] have pointed out that pyrolysis as a technique for the chemical examination of substances logically dates to early man noticing that various materials smelled differently when burned. This olfactory analytical pyrolysis technique was used by forensic examiners well past the turn of this century when laboratory equipment used in the forensic laboratory included little more than microscopes, analytical balances, and Bunsen burners.

The first report of modern forensic analytical pyrolysis appeared in 1962 from the laboratory of Paul Kirk at the University of California, Berkeley,[4] the reported identification of plastics by Py-GC. This technique gained such wide acceptance that, within 15 years, Py-GC was being used in most forensic science laboratories for the examination of trace evidence.[3]

Although the first coupling of pyrolysis and mass spectrometry was reported in 1952 — a study using Py-MS to examine the pyrolysis products of several polymeric materials[5] — the first forensic science use of Py-MS came considerably later. The potential usefulness of Py-MS for the forensic examination of evidential materials was first suggested in 1969,[6] and preliminary tests were performed sometime before 1974.[7] While no data from these tests appear to have been published, the comment was made that the tests had shown considerable promise for the forensic individualization of hair and paint by high resolution Py-MS and that the technique warranted detailed exploration. That detailed exploration of the usefulness of Py-MS to the forensic scientist was not to begin for several years, not until after Py-GC/MS had been used for the examination of evidential materials.

The first use of Py-GC for the examination of evidential materials[4] followed shortly after the first reported couplings of pyrolysis and gas chromatography in 1959.[8-10] The first reported use of Py-GC/MS in a forensic science laboratory, however, did not occur until many years later, no doubt delayed until commercial GC/MS systems became readily available in the early 1970s. A report from the Metropolitan Police Forensic Science Laboratory (MPFSL) in London concerning the characterization of commercial adhesives by Py-GC and infrared spectroscopy described the use of Py-GC/MS for the identification of some of the pyrolysis products.[11]

By the mid-1970s many forensic science laboratories were using GC/MS routinely for the identification and characterization of illicit drugs.[12] Also by this time, some definitive works had been published describing the use of Py-MS for the examination of complex bio-organic materials.[13-15] Apparently working entirely independently, two forensic science laboratories, the Forensic Science Bureau (FSB) of the New Jersey State Police in the U.S.[16] and the MPFSL group in the U.K.,[12,17] published the results of their preliminary efforts applying Py-MS to materials of interest to the forensic scientist. Although the FSB group concentrated upon chemical ionization Py-MS while the MPFSL group primarily used electron impact ionization Py-MS, both groups felt that Py-MS offered considerable potential as a sensitive, rapid comparison technique which might overcome some of the limitations of data interpretation encountered with comparisons by Py-GC.

From the published literature, it is not clear whether the FSB group continued to perform further research with Py-MS or to what extent they used the technique in casework. The MPFSL group, however, clearly did both in the late 1970s. Hughes et al.[18] presented Py-MS mass pyrograms for many common and natural fibers and included a description of a FIT program which they had devised for the computer comparison of two pyrograms. Hickman and Jane,[19] also from MPFSL, published a study of the reproducibility of Py-MS using three different pyrolysis systems. Also from the U.K., in 1979 two Py-MS reports were published by the Home Office Central Research Establishment (HOCRE). One report described a probe for Py-MS and its application to the analysis of commercial polyester fibers,[20] and the other described the use of the probe for Py-MS analysis of acrylic fibers and white alkyd paints.[21] Unfortunately, the circulation of HOCRE reports is restricted to official police laboratories and the material has not been published in the open literature.

By 1979, the use of analytical pyrolysis for the examination of evidential materials had become a substantial portion of the total use of analytical pyrolysis. A review of analytical pyrolysis[22] for the first time included the forensic science use of analytical pyrolysis as a separate section. Also, at about this time appeared, from the MPFSL, a review of the use of analytical pyrolysis techniques in forensic science,[23] and, from the FSB, a review of analytical techniques in forensic science that included analytical pyrolysis techniques.[24] Several years later, also from the FSB, an extensive review of the use of analytical pyrolysis in forensic science appeared.[25]

In addition to the review article cited above,[23] in the first half of the 1980s, the MPFSL group continued to publish applications of Py-MS. The use of Py-MS for the examination of paint coatings from case openers (pry bars) was described in an extensive survey of case openers which compared variations in blade tip shapes, dimensions, and paint formulations in 100 of these tools.[26] An additional report compared the analysis of paint resins by Py-MS, Py-GC, and infrared spectroscopy.[27]

The first analytical pyrolysis report from the Federal Bureau of Investigation (FBI) in the U.S. appeared in 1985, a collaborative study (between the Materials Analysis Unit [MAU] and the Forensic Science Research Group [FSRG], both part of the FBI Laboratory) using Py-GC/MS to identify adhesives used in the construction of improvised explosive devices.[28] That same year, the FBI FSRG also reported using Py-GC/MS for the examination of automotive paints,[29] and the comparison of human hair by Py-GC and Py-GC/MS.[30]

In 1985 the MPFSL group published two additional Py-MS works. Whitehouse et al.,[31] described the results of an interlaboratory trial with Py-MS, and Wheals[32] used examples of Py-GC and Py-MS at MPFSL over the previous decade to illustrate strengths and weaknesses of the two techniques and suggested possible directions for future work.

Following a preliminary report the previous year (1985),[33] in 1986 two scientists from a manufacturer of pyrolysis equipment published a paper describing the examination of photocopy toners using Py-GC/MS.[34] Also in that same year, in response to many inquiries from forensic science laboratories in the U.S. about how to perform capillary column Py-GC, the FBI FSRG published a paper describing how to get started with capillary Py-GC, using as examples data obtained from a Py-GC/MS system.[35]

The MPFSL group published two reports in 1987. One of these reports[36] extended earlier work on adhesives to include newer adhesive products which were commonly used in the U.K. for "do-it-yourself" projects. In the second report,[37] the MPFSL group stepped a bit outside their normal area to apply Py-MS to the examination of natural gums, resins, and waxes from Egyptian mummy cases. In that same year, the FBI FSRG published the results of a survey of the use of analytical pyrolysis by forensic science laboratories in the U.S.[38] Of the 141 forensic science laboratories which reported using analytical pyrolysis, only two laboratories (other than the FBI Laboratory) were using Py-GC/MS and one Py-MS. (The FBI used Py-GC/MS for casework and research, and Py-MS for research.) Two additional studies from the FBI FSRG were reported that year. One report described using Py-GC/MS and Py-GC/MS/MS for the identification of complex organic materials formed during the pyrolysis of human hair.[39] The second described the status of human hair comparisons by Py-GC/MS at the end of the research project on that subject.[40]

The last analytical pyrolysis publication from the FBI FSRG appeared in 1989, a report of the use of Py-GC/MS for the classification of photocopy toners.[41] Also in 1989 the first Py-GC/MS publication appeared from the Forensic Science Laboratory (FSL) of the Government Chemical Laboratories in Perth, Western Australia.[42] This report described a technique to pyrolyze synthetic polymer samples and simultaneously chemically derivatize (methylate) the pyrolysis products prior to analysis by capillary GC and GC/MS, a technique which was referred to as simultaneous pyrolysis methylation capillary gas chromatography (SPM-GC) and SPM-GC/MS.

The FSL group continued publication of investigations using SPM-GC and SPM-GC/MS with two reports in 1991. The first report described the application of the technique to structural determination of alkyd resins.[43] The second report described the use of this technique for the examination of additional classes of materials of interest to the forensic scientist.[44]

Publication by the FSL group continued in 1992 with a review article describing fiber identification by pyrolysis techniques, including the various forms of forensic pyrolysis mass spectrometry.[45]

The latest publication from this group (at the time of this writing) followed the next year with a report describing the characterization of rosin-based commercial resins by SPM-GC/MS.[46] The year 1993 also marked the first forensic pyrolysis-mass spectrometry publications associated with a forensic science laboratory in Japan, the publication of two collaborative works between two groups in the Hyogo Prefecture of Japan (the Forensic Science Laboratory, Hyogo Prefectural Police Headquarters and the Department of Applied Chemistry at the Himeji Institute of Technology). Both works involved the examination of soots by Py-GC/MS as a means for the determination of the main fuel components in accidental fires. The first report dealt with soots formed from various alkylbenzenes[47] and the second with soots from various plastics.[48]

Although not originating from forensic science laboratories, two additional reports appeared in 1993 which should be mentioned as examples of forensic applications of analytical pyrolysis. Four laboratories in Italy collaborated on a study testing the potential for recognition of ancient painting media by Py-GC/MS.[49] While this study primarily showed the potential for identification of the paint binding materials for the purposes of art preservation, the authors commented upon the use of analytical pyrolysis for unmasking forgeries. The second work was a collaboration between two universities and a museum, in and around New York City, and the FOM-Institute for Atomic and Molecular Physics in Amsterdam.[50] This study demonstrated the usefulness of Py-GC and Py-GC/MS for unmasking amber forgeries (imitations of large transparent amber pieces with "inclusions" such as ants, bees, lizards, and mosquitoes).

III. TECHNICAL ASPECTS OF Py-MS

A. Introduction

It seems appropriate at this point to spend some time with the technical aspects of the analytical techniques which are used to perform forensic pyrolysis-mass spectrometry. This discussion will attempt to weave these technical aspects within the framework of the questions that the forensic scientist attempts to answer with the analytical results.

B. Mass Spectrometry

Of the various ionization methods which have been used for the various forms of pyrolysis mass spectrometry, the bulk of the forensic pyrolysis mass spectrometry has utilized 70-eV electron impact ionization. The early Py-MS work performed at the FSB gave examples of the application of isobutane chemical ionization (CI),[16] but no other forensic science laboratory chose to follow this approach for routine casework forensic pyrolysis-mass spectrometry. The FBI FSRG studies on the pyrolysis of human hair did include some methane CI Py-GC/MS[30] and some CI-CAD (collisionally activated dissociation) Py-GC/MS/MS experiments,[39] but these studies were far removed from routine casework.

Py-GC/MS using capillary columns necessitates that rapid scanning of the mass range be used in order to obtain the 10 to 15 scans necessary to adequately define each chromatographic peak. All of the work from Py-GC/MS studies reported from the FBI Laboratory utilized quadrupole mass spectrometers scanned across the range 35 to 350 amu (occasionally a bit higher) at 2 to 3 scans/s.[28–30,35,41] The Py-MS studies from MPFSL began using a magnetic sector mass spectrometer which was incapable of scanning the desired mass range (25 to 250 amu) faster than once every 2 s.[12,17] Because the slow scan rate was a good match to the rate at which the pyrolyzate entered the mass spectrometer with their Py-MS system, later studies with a newer mass spectrometer continued to utilize the slow scan rates and the same mass range.[36,37]

C. Py-GC/MS

Any GC/MS system can be converted to perform Py-GC/MS by the addition of some sort of pyrolysis chamber to the injector end of the GC. In fact, the considerations for making this conversion are the same as those for converting any GC to perform Py-GC. Because a survey of crime laboratories conducted in 1986[38] found that, of the more than 130 laboratories using Py-GC, more than 40 were using narrow-bore capillary column GC (that is, using capillary columns with internal diameters of 0.32 mm or less), and because, in the ensuing years, more laboratories have no doubt switched from packed GC columns to capillary columns, this discussion will focus entirely upon capillary Py-GC/MS. Unless otherwise indicated, references in this discussion to capillary GC will be referring to narrow-bore capillary GC.

As mentioned earlier, in response to many inquiries, this author previously described in detail how to convert a capillary GC system to a Py-GC system.[35] The essence of that report is presented here, but the reader will find much more detail in the original.

1. Py-GC Equipment

This discussion assumes that one of the commercially available GC systems designed to use fused silica open-tubular capillary columns is available. One must be completely proficient in the operation of such an instrument in the split injection mode before attempting the conversion to Py-GC.

For proper operation of the Py-GC system, determine the split ratio (the ratio between the carrier gas flow out the split vent and the flow down the column) for the injector system before and after installation of the pyrolysis interface. The flow out of the split vent can be measured directly by affixing an appropriate gas-flow measuring device onto the split vent. Measuring the flow down the capillary column is a bit more complicated but still not difficult. One can measure the flow directly, either at the end of the column before it has been connected to the detector or out the detector vent after shutting off all other gas flows to the detector. However, these techniques are not very useful for a system which is in operation because they involve disrupting the system. This is especially true of a GC/MS system where disconnecting the column usually means venting the system, and where the high vacuum of the operating system will have a considerable effect upon column flow. A better method for estimating column flow is to measure the retention time of an unretained analyte (such as natural gas, liquid propane gas, or fuel from a disposable cigarette lighter) and to calculate the flow rate. The unretained analyte will elute in one column volume (consider the column to be a cylinder with a volume equal to π times the radius squared times the length). The flow down the column in milliliters per minute will be this volume (in milliliters) divided by the retention time for the unretained analyte in min. If the answer does not come out to be about 1.5 ± 0.8 ml/min for a 0.25-mm i.d. column, an error in measurement, calculation, or consistency of units has probably been made. For running the test samples before and after installation of the pyrolysis interface, use a split ratio between 40 and 100 to 1.

A rather bewildering range of choices confronts one when choosing a capillary column. These include choice of supplier, length, internal diameter, and chemical composition of the stationary phase (as well as phase thickness and whether the phase is coated, bonded, or cross-linked), with four to six choices of each except for phase composition, where there are dozens of choices. However, one can make a single choice which will give, if not the best performance, at least satisfactory performance with all of the samples one will likely encounter as evidence. Such a choice would be a 30-m long by 0.25-mm i.d. fused silica column with a 0.25-μm thick, bonded polysiloxane (containing 5% phenyl) stationary phase. Such a column (a DB-5 column manufactured by J&W Scientific and obtained from Alltech Associates, Deerfield, IL) was used to produce the data shown in Figures 1 through 6. Once one has gained experience with the performance of the GC portion of the system with the types of samples run most often, one may choose to modify this choice to improve some aspect of the pyrograms obtained. A typical temperature program with this column would be 40°C for 2 min, 15°C/min to 250°C with a hold of 5 min. A bit of experience will help one select suitable modifications to this temperature program to improve the results for particular samples.

Several types of pyrolysis units are available today which are easily attached to most GC injection ports. The choices include tube furnaces, Curie-point devices, and resistively heated ribbon or coiled filament pyrolysis probes. At the time of the survey of crime laboratories[38] in 1986, of the 138 laboratories which reported performing Py-GC, 136 used the Pyroprobe (Chemical Data Systems, Oxford, PA), a resistively heated, coiled filament device in which the sample is heated inside a small quartz tube held within the filament coils. For the purpose of this discussion, it will be assumed that the pyrolysis unit is a Pyroprobe.

Every laboratory using capillary GC should have an electronic instrument for detecting minute hydrogen and helium leaks. Such a device is especially useful with a pyrolysis system because several leak-prone connections are added to the injector side of the GC gas-flow pathway. Although the $700 to $900 initial expense may seem high, leaks at the injector are one of the most common causes of poor chromatographic performance, and such a device can save many hours of frustration by quickly pinpointing leak problems. Several suppliers of GC columns and accessories offer such devices.

2. Installation of the Pyrolysis Unit

To perform pyrolysis capillary GC, a heated pyrolysis interface chamber must be installed on the injector of the GC (except for certain GCs which allow for direct insertion of the Pyroprobe into a modified capillary injector). Because addition of the interface will degrade GC performance slightly if done properly, and severely if done improperly, prior to installation it is desirable to establish "benchmarks" against which to measure subsequent performance.

Figure 1 shows the total ion current profile (TICP) obtained from the analysis of the Grob test mix which can be used for testing the entire injector-column-detector system of a GC or GC/MS system. The peak shapes and relative peak heights of the test mix components indicate whether the system will perform satisfactorily with various classes of acidic, basic, polar, and nonpolar chemicals.[51]

Because the components in the Grob test mix are relatively low boiling, and because the interface-injector assembly may have one or more "cold spots" depending upon the injector design, an additional test mixture to check for possible "cold trapping" of less volatile components should be run for later comparison. Figure 2A shows the TICP obtained from an n-alkane mix (C_{20}, C_{22}, C_{24}, and C_{26}) prior to installation of the pyrolysis interface. It is of critical importance to evaluate the performance

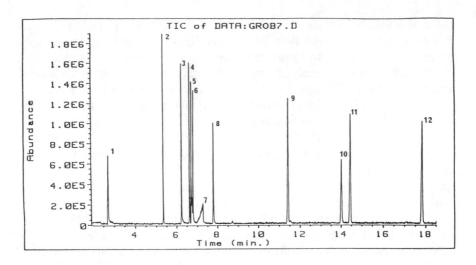

Figure 1.

Capillary GC/MS analysis of the Grob test mix: (1) 2,3-butanediol; (2) decane; (3) 1-octanol; (4) undecane; (5) nonanal; (6) 2,6-dimethylphenol; (7) 2-ethylhexanoic acid; (8) 2,6-dimethylaniline; (9) methyl decanoate; (10) dicyclohexylamine; (11) methyl undecanoate; and (12) methyl dodecanoate. (From Munson, T. O., *Crime Lab. Digest*, 13, 82, 1986. With permission.)

of a GC/MS system in this fashion prior to the installation of a pyrolysis unit. If the GC/MS system will not perform satisfactorily at this point, the situation will not likely improve after conversion for pyrolysis. The cited test mixes are available from several commercial suppliers of GC supplies.

The function of the interface chamber is to provide a heated space in which the sample can be pyrolyzed and from which the resulting pyrolyzate can be transferred to the GC injector. The interface chamber is connected to the GC injector via a needle assembly provided by the manufacturer of the Pyroprobe. It is necessary to have the needle assembly for the particular GC injector to be used because the assembly is designed to replace the GC septum nut and must be the correct size. When a sample is pyrolyzed, the pyrolyzate is swept by carrier gas from the chamber into the injector via the needle assembly. Upon reaching the injector, the pyrolyzate proceeds exactly as if it were a vaporized sample which had been injected into the split injector with a syringe. The installation procedure consists of three important steps: installing the needle assembly, mounting the interface chamber upon the needle assembly, and establishing the carrier gas flow.

The installation of the needle assembly is easily accomplished once the correct size for the injector has been obtained. The GC carrier gas is turned off, the pressure is allowed to decrease, and the injector septum nut is

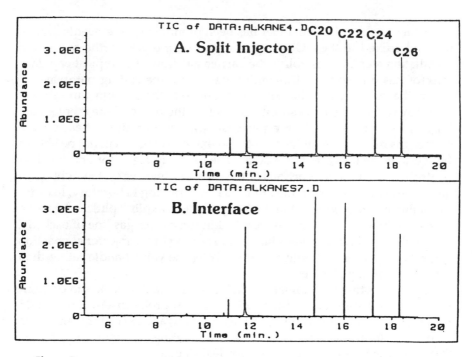

Figure 2.
Capillary GC/MS analysis of the alkane test mix injected into (A) split injector and (B) pyrolysis interface. (From Munson, T. O., *Crime Lab. Digest*, 13, 82, 1986. With permission.)

removed and replaced with the needle assembly. The needle assembly will probably come with a septum in it, but eventually this septum will need to be replaced, and there is a problem awaiting the unwary. In order that it might fit into the capillary injector, the needle on the assembly has a small diameter and is very easily plugged by being pushed through a silicone-rubber septum. Once in this condition, the needle is unplugged only with great difficulty. Prior to replacing the septum in the needle assembly, a hole larger than the needle shaft should be cored through the septum (a gas-tight seal is not required around the needle shaft).

Mounting the interface chamber is easily accomplished if a new Pyroprobe with the appropriate mounting brackets has been acquired to fit an existing capillary GC. If this is not the case, the Pyroprobe supplier can furnish the appropriate mounting brackets. It is important to mount the interface chamber solidly to the GC so that inserting and removing the Pyroprobe from the interface chamber (which involves tightening and loosening the retaining ring containing the silicone-rubber seal around the probe) does not exert torque on the interface chamber. This process can stress the seal where the interface chamber connects to the needle assembly and cause leakage of the carrier gas.

When the interface chamber has been mounted on the needle assembly and affixed to the GC with the appropriate brackets, the remaining installation step is to establish the carrier gas flow. The typical capillary injector has one source of incoming gas and three exiting pathways — down the column, out the split vent, and out the septum purge vent. Disconnect the carrier gas supply just before the point where it enters the injector and connect it to the pyrolysis interface chamber. Seal off the carrier gas inlet fitting on the injector so that carrier gas will not be able to flow out of the injector at this point. The pressure regulator and/or flow controller and all other portions of the capillary injector system will operate in the usual fashion. In the FBI FSRG GC Training Laboratory, in order to facilitate rapid conversion from pyrolysis to split/splitless injection modes, a three-way valve was installed in the carrier gas line of each GC so that the gas flow pathway change and sealing of the injector connection could be accomplished with a single turn of the valve handle rather than several plumbing changes.

Prior to testing the capillary Py-GC system, the final task is to pressurize the interface-injector system with the Pyroprobe inserted (about 15 psig for a system using a 30-m capillary column) and check for leaks. Then heat the interface-injector assembly to operating temperature (at least 150°C for the interface) and check for leaks again.

3. Performance Validation

After the system has been shown to be leak-free, test the performance of the interface-injector assembly by running the test samples again. Remove the Pyroprobe, seal the interface chamber with a septum, establish the same split ratio used for the earlier runs of the test mixes, and check around the septum for leaks. Then rerun each of the test mixtures by injecting them directly into the interface chamber.

Figure 3 shows the TICP obtained for the Grob test mix after installation of the pyrolysis interface (3B) compared to that obtained prior to the installation (3A). If everything went well, the before and after TICPs should be nearly the same as far as peak shape, retention time, and peak heights (as those shown in Figures 3A and B are). If the peak heights and/ or peak shapes are poor, the interface-injector assembly may have a leak. If the alkane peaks in the Grob test mix have good shapes and responses, but some of the others do not, the hot stainless steel in the pyrolysis interface may be degrading some of the compounds. Repeated pyrolysis of some fairly large samples seems to improve this situation, perhaps by coating or deactivating reactive sites.

Figure 2 shows the TICPs obtained for the alkane test mix with the GC/MS system before (2A) and after (2B) installation of the interface. The regular stepwise decrease in the C_{22}, C_{24}, and C_{26} alkanes when injected

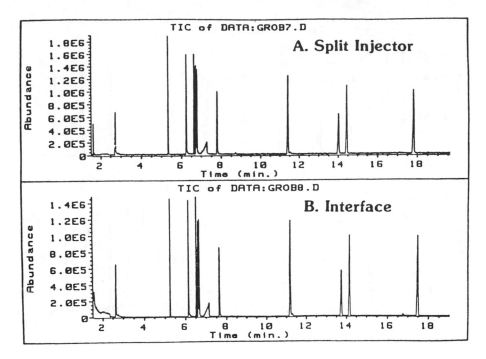

Figure 3.
Capillary GC/MS analysis of the Grob test mix injected into (A) split injector and (B) pyrolysis interface. (From Munson, T. O., *Crime Lab. Digest*, 13, 82, 1986. With permission.)

into the interface-injector assembly suggests a small amount of "cold-trapping" and, hence, slightly degraded performance relative to before installation of the pyrolysis interface, but not enough to adversely affect the analysis of typical forensic samples.

With a pyrolysis system which can pass the tests enumerated above, it should be possible to generate very reproducible pyrograms. If the GC is operating properly, retention times for capillary runs should reproduce within ±1 s from one run to the next.

To avoid two problems which can lead to nonreproducible pyrograms, the size of the sample pyrolyzed should be kept small, less than 100 µg, even when large amounts of sample are available. The thermal energy put into the sample during pyrolysis causes bond breakage and the formation of radicals which are fragments of the original polymer molecules. If the concentration of the radicals is high enough, some of the components of the pyrolyzate will be formed by radical-radical interactions rather than by intramolecular rearrangements of the radical molecules, thus causing the composition (the number and relative amounts of the chemical components) of the pyrolyzate to vary with the sample size. In such a case, subsequent pyrolysis of a second portion of the sample will produce a

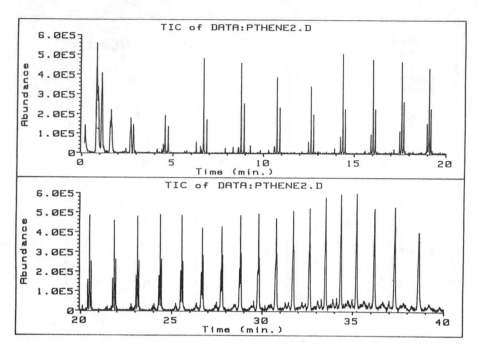

Figure 4.
Capillary Py-GC/MS analysis of polyethylene. (From Munson, T. O., *Crime Lab. Digest*, 13, 82, 1986. With permission.)

pyrogram with noticeable differences from the first, a highly undesirable situation when one is comparing various samples for similarity.

Another problem related to sample size can occur if the sample is so large that it is not completely consumed during the pyrolysis period. Many samples are uniform mixtures of components, such as an evenly layered paint chip or a fragment of tape consisting of backing coated with an adhesive layer. Unless such samples are completely consumed during pyrolysis, differing relative amounts of the various components of the mixture may be produced each time the sample is pyrolyzed, resulting in differing pyrograms. To test whether a sample has been pyrolyzed to completion, perform a second pyrolysis without removing the probe; obviously, if no peaks appear, no residual sample was present (an assumption underlying this test is that an appropriate system blank has been demonstrated — that is, pyrolysis of an empty quartz sample tube has resulted in no peaks in the pyrogram). If a sample material being pyrolyzed is, in fact, randomly heterogeneous, one should expect the pyrograms for repetitive analyses to differ.

Obtaining an appropriate pyrogram from the pyrolysis of a well-characterized polymeric material constitutes a suitable test for satisfactory performance of the entire pyrolysis system. Figure 4 presents a pyrogram for low-density polyethylene. This material tests the system well for the

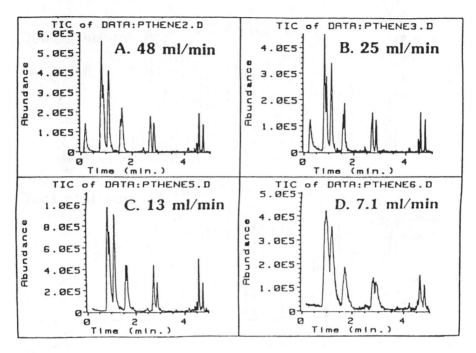

Figure 5.
Comparison of the portions of the capillary Py-GC/MS polyethylene pyrograms with interface carrier gas flows of: (A) 48 ml/min; (B) 25 ml/min; (C) 13 ml/min; and (D) 7.1 ml/min. (From Munson, T. O., *Crime Lab. Digest*, 13, 82, 1986. With permission.)

entire volatility range generally found in pyrograms of evidential materials and will be used here to demonstrate potential problems with two operating parameters of the pyrolysis system — carrier gas-flow rate through the interface and temperature of the interface.

Increasing or decreasing the carrier gas flow through the pyrolysis interface at a given head pressure will increase or decrease the split ratio and concomitantly decrease or increase the amount of the sample pyrolyzate which goes onto the GC column. When routinely working with very small samples, one may wish to use a split ratio as small as possible to maximize the peak heights in the pyrogram. If the split ratio (and hence, the carrier gas flow) is decreased too much, the pyrolyzate will not be swept through the capillary injector rapidly enough to get the components added to the GC column as narrow bands, and resolution will be adversely affected. Figure 5 shows expanded views of the early portion of the polyethylene pyrogram using different carrier gas flows: (A), 48 ml/min, a 38:1 split ratio; (B), 25 ml/min, a 20:1 split ratio; (C), 13 ml/min, a 10:1 split ratio; and (D), 7.1 ml/min, a 5.6:1 split ratio. The sample size was decreased along with the split ratio in an attempt to keep the peak heights comparable. The resolution of those components which elute from the

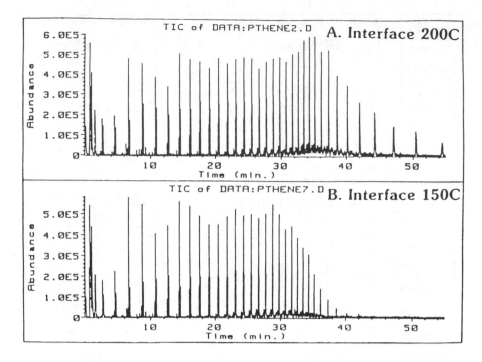

Figure 6.
Comparison of the capillary Py-GC/MS pyrograms of polyethylene at interface temperatures of (A) 200°C and (B) 150°C. (From Munson, T. O., *Crime Lab. Digest*, 13, 82, 1986. With permission.)

column prior to 5 min is considerably diminished by the lowest flow rate. For this system, a split ratio of 10:1 appears to give satisfactory performance; however, anyone planning to use a split ratio of less than 20:1 should perform this experiment to determine the characteristics of the interface-injector system being used. This loss of resolution was not observed for the components which eluted from the column after 5 min, probably due to a band-narrowing effect resulting from condensation of the components at the head of the column at the start temperature. If the information in the early portion of the chromatogram is of no interest, the splitless injection technique might be useful for very small samples.

Some analysts prefer to have the pyrolysis interface at 150 rather than 200°C because the probe seal area stays cooler, perhaps reducing "bleed" from the silicone-rubber seal as well as prolonging the life of the seal and causing fewer burned fingers. For most forensic samples, the lower temperature probably serves quite well, but one should be aware of a potential problem. Figure 6 shows the pyrogram for polyethylene with the interface at 200°C (6A) compared to the pyrogram for polyethylene with the interface at 150°C (6B). If the main interest is in the late eluting materials, the lower interface temperature may not be suitable due to the rather serious "cold trapping" of these components of the pyrolyzate.

D. Py-MS

As was mentioned earlier in this chapter, with Py-MS, the sample pyrolyzate is either generated at, or passes directly into, the ionization source of the mass spectrometer without an intervening separation process such as takes place in Py-GC/MS. While at first blush this process may seem much simpler than Py-GC/MS and therefore more likely to lead to reproducible results, it turns out that the several options for establishing Py-MS on an existing mass spectrometry system can lead to quite different results for the Py-MS analysis of various types of materials.

This chapter will not attempt to provide an overview of the many Py-MS techniques that can be found in other publications,[13–15,22,52] but will be limited to the Py-MS techniques which are being or have been used in forensic science laboratories. Other than the previously mentioned comments about Py-MS from De Forest[6,7] which gave few details, four forensic science laboratories have been identified which have used or are using Py-MS: the Forensic Science Bureau (FSB) of the NJ State Police and the Forensic Science Research Group (FSRG) of the FBI Laboratory in the U.S.; and the Metropolitan Police Forensic Science Laboratory (MPFSL) and the Home Office Central Research Establishment (HOCRE) in the U.K.

As alluded to earlier, the HOCRE group used a Py-MS probe for the analysis of fibers and white alkyd paints.[20,21] Unfortunately, this author was not able to obtain a copy of the first reference wherein the technical details about the probe design were presented, and no details about the probe were included in the second publication. Fortunately, however, the HOCRE group loaned this probe to the MPFSL group for a study of three Py-MS systems. In the report of this study,[19] the probe was described as being fabricated by VG to fit in place of the direct insertion probe of the mass spectrometer, incorporating a Curie-point pyrolyzer (15-s pyrolysis at 610°C), and passing the pyrolysis products directly into the ionization source.

The Py-MS system at the FBI FSRG consisted of a coiled filament-type pyrolysis probe which was custom-manufactured by the makers of the Pyroprobe to exactly replace the direct insertion probe on the Finnigan TSQ45. This probe was designed so that pyrolysis products would travel only a few millimeters before entering the source ion volume. The intention was to examine human hair pyrolyzate by Py-MS and Py-MS/MS for larger fragments than would pass down the GC column during previous Py-GC/MS and Py-GC/MS/MS experiments.[39] A few preliminary experiments testing the operation of this probe were performed, but nothing further.

Direct-insertion Py-MS probes which pass the pyrolyzate directly into the ionization source of the mass spectrometer can cause the deleterious buildup of organic deposits in the mass spectrometer unless the system has been specially designed for this process.[52] The FSB and the MPFSL groups both reported arrangements where the pyrolyzate was generated at some considerable distance from the ionization source of the mass spectrometer.

In the FSB work,[16] the heated interface chamber of a coiled filament type pyrolysis probe was connected to the batch inlet of the mass spectrometer. The samples were pyrolyzed into the evacuated batch inlet with the valve open. The pyrolyzate was held in the batch inlet for 60 s before opening the analyzer valve and allowing the pyrolyzate to leak through a small orifice into the ionization source of the mass spectrometer. The mass spectrometer was operated in the isobutane CI mode. The simple CI spectra obtained from the analysis of various paints and fibers contained many ions which were associated with the expected thermal degradation products of the polymers analyzed.

The MPFSL Py-MS work began with the connection of a Curie-point pyrolyzer to a GC/MS system used primarily for drug analysis.[17] In order to facilitate rapid switching between Py-MS and GC/MS, the pyrolyzer was connected via an empty glass chromatography column (45 cm × 2 mm i.d.) to the same heated glass line and jet separator as were used to transfer the GC eluate to the mass spectrometer. The empty GC column was held at 200°C in the GC oven and the pyrolyzate swept down the column with a helium flow of about 6 ml/min. Helium makeup gas flow as added at the end of the column to increase the total flow to the jet separator to 15 ml/min.

Later on, after 1979, the MPFSL primarily used a Pyroprobe coiled filament pyrolyzer mounted in their Py-MS system,[53] as shown in Figure 7. With this setup, there was a shorter, wider tube for the pyrolyzate to expand into (a 23-cm × 7-mm i.d. tube connected to a 3-cm × 3-mm i.d. tube). Because the Pyroprobe was fully inserted into the larger tube, the pyrolyzate expanded into only about 9 cm of the larger tube. The pyrolyzate was swept down the tube assembly with helium at 6 ml/min and no makeup gas was added.

The expansion space in both of the MPFSL Py-MS configurations was critical in order to allow the pyrolyzate, which was generated in a few seconds, to expand out and enter the mass spectrometer for a minute or more. This condition allowed multiple scanning of the magnetic sector mass spectrometer which was used for their Py-MS analyses. The total cycle time for scanning the usual mass range used (25 to 250 amu) was about 2 s. They found it highly desirable to use mass pyrograms which were averaged from at least 6 and, in some cases, more than 12 scans.[53]

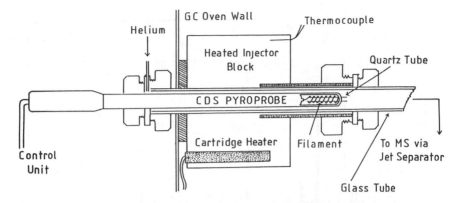

Figure 7.
Diagram showing how the CDS Pyroprobe pyrolyzer was mounted for Py-MS at the Metropolitan Police Forensic Science Laboratory. (From Wheals, B. B., in *Proceedings of the Int. Symp. on the Analysis and Identification of Polymers*, Federal Bureau of Investigation, Quantico, VA, 1985. With permission.)

After obtaining the Pyroprobe, the MPFSL took the opportunity to compare three Py-MS configurations using the two described immediately above and the VG direct insertion probe borrowed from the HOCRE group.[19] Replicate samples of three polymer types were analyzed over time on each of the three systems. The polymers ranged from easy to difficult: a simple polyvinyltoluene-polystyrene wire coating; a yellow acrylic paint known from Py-GC studies to pyrolyze reproducibly; and a white alkyd paint known from Py-GC studies to pyrolyze in a variable manner. A FIT factor, developed earlier,[12] was used for measuring the similarity of pairs of mass pyrograms, with a perfect match in ion masses and relative intensities giving a FIT factor of 1000. The comparison of mass pyrograms from same-day analyses of the acrylic paint gave average FIT factors of 999.8, 998, and 995 for the Curie-point, Pyroprobe, and VG probe systems, respectively. The comparison of mass pyrograms for the acrylic paint analyzed on different days over a 6-week period gave average FIT factors of 996, 996, and 991, respectively. Figure 8 shows typical mass pyrograms for this yellow acrylic paint analyzed on these three Py-MS systems. Although the mass pyrograms appear quite similar, the FIT factors for the intercomparisons of the three mass pyrograms shown are less than 900.

The comparison of mass pyrograms from same-day analyses of the alkyd paint gave average FIT factors of 996, 995, and 994 for the Curie-point, Pyroprobe, and VG probe systems, respectively. The comparison of mass pyrograms for the alkyd paint analyzed on different days over a 6-week period gave average FIT factors for 994, 978, and 976, respectively. Figure 9 shows typical mass pyrograms for this white alkyd paint ana-

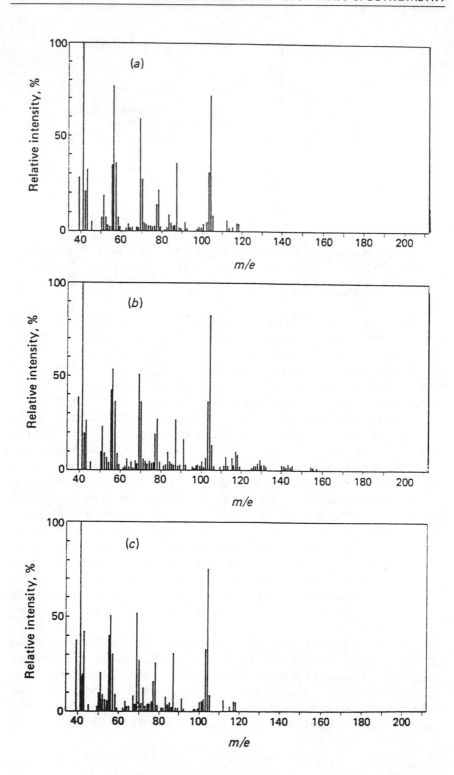

Figure 8.
Mass pyrograms for yellow paint. (a) Curie-point pyrolysis into glass column; (b) CDS Pyroprobe into glass column; and (c) VG pyrolysis probe. (From Hickman, D. A. and Jane, I., *Analyst*, 104, 334, 1979. With permission.)

lyzed on these three Py-MS systems. Not only was the reproducibility poorer for the same-day comparisons, as well as over extended periods, for this paint compared to the acrylic paint, one can see that the composition of the mass pyrograms obtained was much more dependent upon the type of Py-MS system utilized than was the case for the yellow acrylic paint.

Analysis of intracompound ion abundance ratios of the mass pyrograms obtained during this study yielded the important observation that the mass spectrometric detection of the components of the pyrolyzate seemed only to introduce small errors, no more than 2%. The major source of irreproducibility in the Py-MS analysis appeared to be in the pyrolysis process. This same conclusion was reached in a later study[31] which compared the results obtained when six laboratories examined identical samples by Py-MS using various system types. Two of the samples were common evidential materials — a polystyrene-poly(vinyl toluene) blend and an alkyd resin typical of those used for gloss paints.

E.   Data Analysis

The forensic scientist dealing with the types of evidential materials which would be examined by forensic pyrolysis would be trying to answer three questions which have been succinctly stated[32] as follows: (1) what are the samples; (2) if more than one sample was examined, were any analytically identical; and (3) if they were identical, how significant is this?

Early on, Py-GC was widely used (as it is today) for comparing bits of trace evidence for commonality (question number 2). In other words, was material A associated with a crime scene analytically the same as material B associated with a suspect (or another crime scene)? While this question was rather easily answered by Py-GC — either the two materials appeared the same analytically or they did not — what significance to attach to an observed analytical match was not at all clear (question number 3). One could not convincingly claim in a court of law that the chip of white paint taken from the trousers of a suspect "linked" him to the crime scene because the Py-GC profile (pyrogram) of the paint chip matched the pyrogram of white paint at the crime scene unless one knew how many such matches were likely if pyrograms of all similar white paints were compared to the evidential paint. One clearly needed searchable reference libraries containing pyrograms obtained under "standard" pyrolysis conditions of the various different types of materials which commonly were found as trace evidence. Because Py-GC analysis using packed GC col-

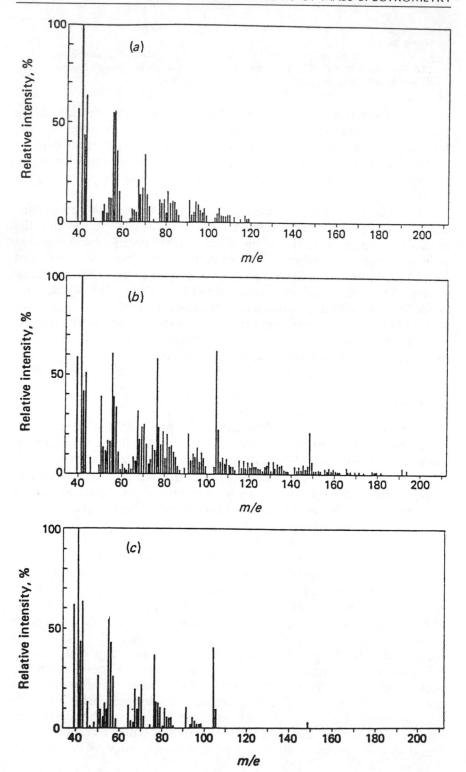

Figure 9.
Mass pyrograms for white alkyd paint. (a) Curie-point pyrolysis into glass column;
(b) CDS Pyroprobe into glass column; and (c) VG pyrolysis probe. (From Hickman,
D. A. and Jane, I., *Analyst*, 104, 334, 1979. With permission.)

umns did not provide reproducible results in interlaboratory comparisons, and because pyrograms did not lend themselves to automated data
handling techniques, the development of such libraries did not seem very
feasible. In addition, no comparison technique more formal than "eye-
balling" the chromatogram had been developed for comparing pyrograms
for similarities and for judging how similar a particular pair of pyrograms
might be.

Py-MS appeared to offer the potential for addressing question number
3. In addition to offering an analysis which took a few minutes to compare
two substances rather than the hour or more required by Py-GC, Py-MS
analysis provided data more amenable to computer processing.

At the MPFSL over a period of about 6 years, Py-MS essentially
replaced Py-GC for the examination of trace evidence by analytical pyrolysis. This trend was largely due to the accumulation of a computer-
searchable library of mass pyrograms which could be used not only to
address the what-is-it question, but also to render some opinion about
how significant analytical similarities were.[32]

The introduction of fused silica capillary GC columns in the late 1970s
enhanced both the discriminating power and the reproducibility of Py-
GC, but still left the problem of how to compare the pyrograms in some
automated and/or quantitative fashion. Two Py-GC/MS publications from
the FBI FSRG offered an approach to this data handling problem.[28,29]

Modern GC/MS data systems have searchable libraries of reference
spectra. In addition, the software allows one to create new libraries consisting of spectra which have been generated on the GC/MS system. The
software also allows one to manipulate spectra, for instance, subtracting
one spectrum from another, or summing spectra to create an averaged
spectrum. By summing all of the spectra from a GC/MS analysis, one can
create a spectrum which represents a weighted average of the mass spectra of all of the components in the GC/MS profile. This spectrum takes on
a special significance when applied to a Py-GC/MS data file because, in
this instance, the summed spectrum is very much akin to the mass pyrogram
of a Py-MS analysis.

This data manipulation technique was tested with Py-GC/MS data
first with 39 automotive paints[29] and then with 91 commercially available
adhesive products.[28] Figure 10 presents the Py-GC/MS total ion current
profile derived from "Black Magic" brand adhesive with some of the
major components identified. The computer was instructed to compose a
spectrum using scans 100 to 1500 and to make a background correction
using scans 10 to 50. The resulting composite spectrum shown in Figure

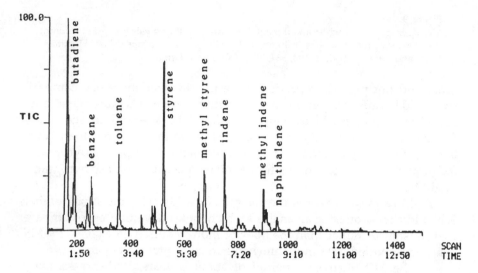

Figure 10.
Total ion current profile from the Py-GC/MS analysis of "Black Magic" adhesive
with some components identified. (From Bakowski, N. L., Bender, E. C., and Munson,
T. O., *J. Anal. Appl. Pyrol.*, 8, 483, 1985. With permission.)

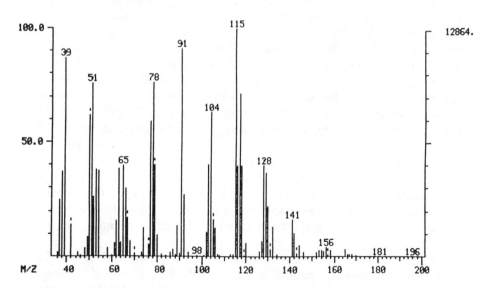

Figure 11.
Composite mass spectrum derived from the Py-GC/MS profile for "Black Magic"
brand adhesive shown in Figure 10. (From Bakowski, N. L., Bender, E. C., and
Munson, T. O., *J. Anal. Appl. Pyrol.*, 8, 483, 1985. With permission.)

11 has all of the ions from the mass spectra of all the components in the pyrogram: 39 amu from butadiene; 78 amu from benzene; 91 and 92 amu from toluene; 78 and 104 amu from styrene; 115 and 116 amu from indene; 115, 117, and 118 amu from methylstyrene; 115, 129, and 130 amu from methylindene; and 128 amu from naphthalene. The relative abundances of the ions in the composite mass spectrum reflect the relative concentrations of all of the components in the pyrolyzate. This composite spectrum was then added to a library file as the entry representing this particular adhesive.

Although these two studies represented only preliminary evaluations of this data handling technique, the results looked quite promising. After reviewing the first presentation[54] of this approach with automotive paints, Wheals of MPFSL cited several potential advantages,[32] among them: (1) because each eluting compound yields a mass spectrum, variation in GC peak retention time would be less of a problem; (2) the total ion current profile from each analysis provides a record similar to a Py-GC pyrogram for comparison; and (3) the summed spectrum, which is akin to a Py-MS mass pyrogram, can be used as a "fingerprint" for computerized library storage and comparison.

IV. EXAMINATION OF EVIDENTIAL MATERIALS

A. Introduction

Paints of various types comprise the most common samples which are examined by forensic pyrolysis. In a 1985 summary of the use of pyrolytic methods at MPFSL,[32] Wheals estimated that in 1973, when Py-GC was the only pyrolytic method being used at MPFSL, 72% of the samples were paints of various types (vehicle paints, 35%; decorative gloss paints, 33%; and tool paints, 4%). No other sample class made up more than 7%. In 1983, when 85% of the pyrolysis at MPFSL was Py-MS, 58% of the samples were paints (vehicle, 28%; decorative gloss, 25%; tool paints, 4%; and emulsion paints, 3%). Adhesives and plastics (including plastic tapes) made up the bulk of the remaining samples (14 and 15%, respectively).

In a 1987 survey of the use of analytical pyrolysis by forensic science laboratories in the U.S.,[38] 105 of the 141 laboratories which were performing analytical pyrolysis responded to the question about types of samples examined. The three most common responses were paint, 96 laboratories; plastics (a summation of polymer and plastic), 53; and fibers, 40. The other reported uses were as follows: rubber, 22; petroleum products, 12 (including grease, 5; asphalt/tar, 2; oils, 3; and waxes, 2); and 5 laboratories reporting other specific uses (dyes, 1; gum, 1; and adhesives, 3). At the time of this survey, the bulk of the casework-associated forensic pyrolysis mass spectrometry in the U.S. was Py-GC/MS being performed in the MAU of the FBI Laboratory. The most common sample being examined at the FBI MAU was automotive paint, followed, as a distant second, by tapes and adhesives recovered from improvised explosive devices.

B. Paints

In their study of case openers, Castle et al.,[26] examined the paint from 52 of the tools by Py-MS. Using Py-MS alone, the 52 paints could be classified into seven groups based upon resin type, and there were enough individual differences within the seven groups that 27 unique mass pyrograms were obtained. Figure 12 displays mass pyrograms of three alkyd paints: (1) unmodified, characterized by phthalic anhydride (m/z 104, 76, 148, 50); (2) styrene modified (additional ions at m/z 104, 103, 91, 51, 78); and (3) vinyl toluene modified (additional ions at m/z 117, 118, 91, 115).

At MPFSL, Py-MS has proven to be a very useful system for the comparison of automotive paints, with various vehicle paint systems giving quite characteristic pyrograms.[53] Figure 13 shows mass pyrograms from three types of vehicle paints.

An interesting approach using Py-GC/MS to construct a computer-searchable library of reference spectra for the analysis of automotive paints was mentioned earlier.[29] Figure 14 shows Py-GC/MS pyrograms for two blue automotive paints. When the spectra which comprised the total ion current profile for paint A were combined into a composite spectrum (as described in Section III.E) and the composite spectrum searched against a library made up of composite spectra from 39 automotive paints, the computer correctly matched the blue paint with a match purity of 976. The next closest match from among the other six blue paints in the library was a purity of 907, but two brown paints were fairly close matches at 962 and 932. The library composite spectrum for blue paint B in Figure 14 had a match purity of 760 compared to paint A.

C. Adhesives

In a Py-MS study of 94 household adhesives,[36] it was amply illustrated that mass pyrograms can be used to identify main pyrolysis products in addition to being used as "fingerprints" for comparison. Figure 15 shows the mass pyrograms for four adhesive systems which employ rubber-like resin as bases: (1) polyisoprene; (2) styrene-butadiene; (3) nitrile; and (4) neoprene. These mass pyrograms represent difficult examples because,

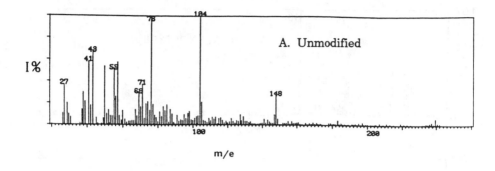

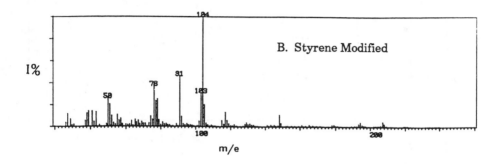

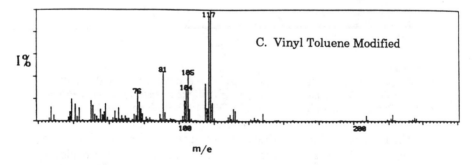

Figure 12.
Py-MS mass pyrograms of alkyd paints from case openers pyrolyzed at 700°C for 20 s. (From Castle, D. A., Curry, C. J., and Russell, L. W., *Forensic Sci. Int.*, 24, 285, 1984. With permission.)

unlike simple adhesive systems which pyrolyze to yield monomers and oligomers, the mass pyrograms shown in Figure 15 are quite complex due to the incorporation of compounding ingredients into the rubber-like resin bases. Nonetheless, in Figure 15A, isoprene (m/z 67, 68), isoprene dimer (m/z 136), and dipentene (m/z 68, 93) can be seen, and in Figure 15B, ions associated with styrene (m/z 78, 103, 104) and butadiene (m/z 54) can be

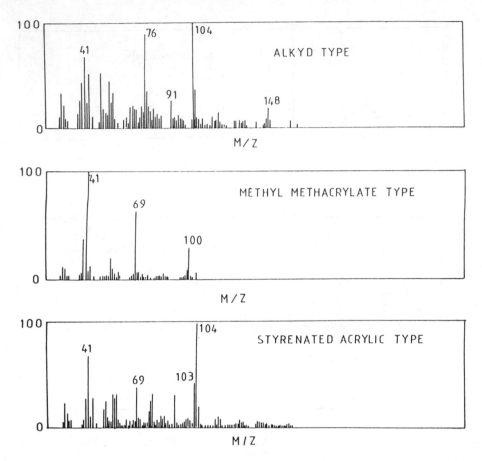

Figure 13.
Py-MS mass pyrograms of vehicle paints pyrolyzed at 700°C for 20 s. (From Wheals, B. B., in *Proceedings of the Int. Symp. on the Analysis and Identification of Polymers*, Federal Bureau of Investigation, Quantico, VA, 1985. With permission.)

seen. Nitrile rubber-based adhesive systems employ a wide range of modifying copolymers and additives, and, consequently, provide quite complex mass pyrograms such as the one in Figure 15C. Although individual components cannot be identified with certainty, the overall pattern is characteristic of this type of adhesive system. The mass pyrogram shown in Figure 15D from the neoprene-based adhesive is also quite complex, but one can see ions from the monomer (chloroprene, m/z 88) and from additives commonly found in this class adhesive — methyl abietate, derived from the rosin esters (m/z 239, 197, 195), *tert*-butylphenols (m/z 163, 179, 135, 149), and phthalate ester plasticizers (m/z 149).

Samples of 91 commercially available adhesives were analyzed by Py-GC/MS, and composite spectra used to build a searchable library, as

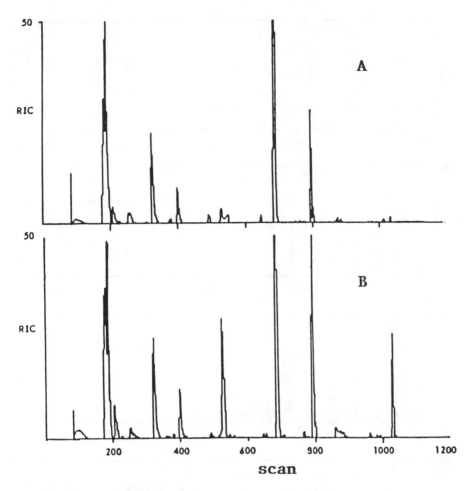

Figure 14.
Py-GC/MS pyrograms of two blue automotive paints. (From McMinn, D. G., Carlson, T. L., and Munson, T. O., *J. Forensic Sci.*, 30, 1064, 1985. With permission.)

described above.[28] Figures 10 and 11, and the associated discussion in Section III.E, provide examples of this data handling technique. As had been observed at the MPFSL in a previous Py-GC and infrared spectroscopy study of adhesives,[11] most of the adhesives fell into a few general classes depending upon the type of adhesive base. Figure 16 shows examples of the total ion current profiles (TICPs) from the Py-GC/MS analysis of six different types of adhesives. Except for the cyanoacrylic-based adhesives ("super-glues"), different brands of adhesive which fell into the same class could be differentiated based upon the presence or absence of minor components and/or differences in amounts of certain components. These characteristic differences were quite reproducible. The eight brands

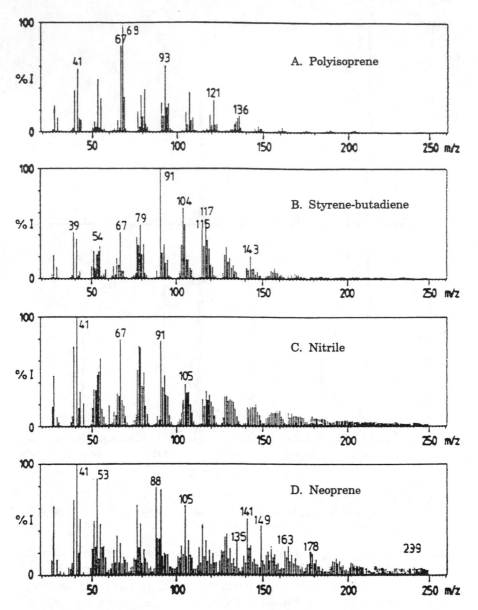

Figure 15.
Py-MS mass pyrograms from four rubber-based adhesives. (From Curry, C. J., *J. Anal. Appl. Pyrol.*, 11, 213, 1987. With permission.)

of super-glues, however, gave TICPs with only very minor differences which did not seem sufficient as distinguishing features. In the Py-MS adhesive study cited above,[36] the mass pyrograms obtained for the different brands of super-glues studied also were indistinguishable.

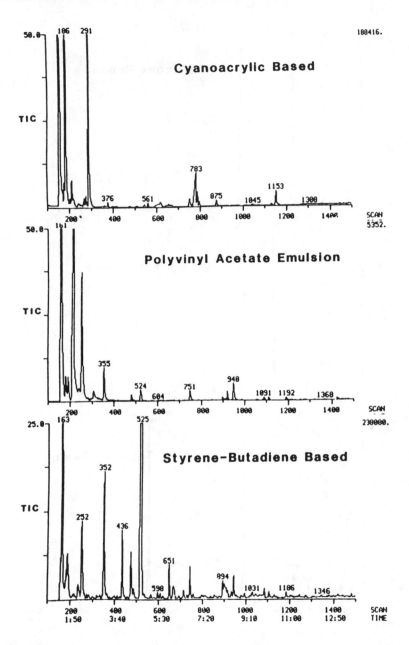

Figure 16.
Total ion current profiles from the Py-GC/MS analysis of six types of adhesives. (From Bakowski, N. L., Bender, E. C., and Munson, T. O., *J. Anal. Appl. Pyrol.*, 8, 483, 1985. With permission.)

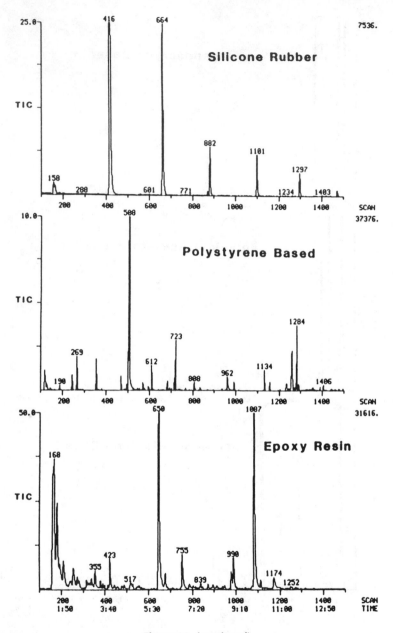

Figure 16. (continued)

Whether or not a searchable library for adhesives, such as that cited above, will still perform usefully with hundreds of entries remains to be seen, but, with 91 entries, the results were quite satisfactory. Five improvised explosive devices (IEDs) were constructed using 24 adhesive products from the library collection. After the IEDs were individually detonated in an underground bunker filled with sandbags, 19 adhesive residues representing 13 of the initial 24 adhesive products were recovered. When these residues were analyzed by Py-GC/MS and the composite spectra searched against the adhesives library, in 7 of the 13 samples, the highest ranked adhesive was the correct match, and, in all cases, the correct match was among the top five reported. In this case, the primary purpose for the searchable library was to help the analyst quickly winnow through the entire library collection by searching the composite spectra against the library to identify the best candidates for further comparison using the TICPs. The correct matches were quite clear from the TICPs (except for the super-glues).

D. Fibers

It appears that pyrolysis mass spectrometry (in any form) has not been used routinely for the analysis of fibers, at least as of the late 1980s. As late as 1987, the FBI Laboratory, the group which performed most of the routine forensic analytical pyrolysis mass spectrometry in the U.S., did not use Py-GC/MS at all for fibers (nor did they routinely use Py-GC), relying almost entirely upon microscopic comparisons and microchemical tests. The MPFSL, the group which performed most of the routine forensic analytical pyrolysis in the U.K., did not report fibers among the types of samples being examined as evidence by Py-MS in 1983, although some may have been included in the plastics/tapes category which made up 15% of the casework.[32]

It is clear that Py-GC, especially using capillary columns, is a very useful technique for the analysis of fibers.[45] In the survey mentioned earlier,[38] of the 141 forensic science laboratories in the U.S. which reported using analytical pyrolysis, 40 of the 105 laboratories which responded to the sample type question used the Py-GC for the examination of fibers. Figure 17 shows a capillary GC pyrogram of a polyacrylonitrile fiber (Creslan 61) which exhibits the pattern of dimers, trimers, and tetramers produced from pyrolysis of the homopolymer of acrylonitrile. If, as in this case, one has the additional information provided by Py-GC/MS about some of the minor peaks, one can speculate about copolymers in the fiber. For instance, peak 1 may be acetic acid from vinyl acetate, and peaks 2 and 3 might have come from vinyl pyridine and or pyrrolidone.

Although the capabilities of both techniques have changed since 1978, a report from MPFSL in that year found Py-MS to have more discriminating power than infrared spectroscopy (IR) for a collection of nylon fibers.[18] Figure 18 shows the mass pyrograms for four types of nylon fibers which were discriminated by Py-MS. Of the eight types of nylon examined, two

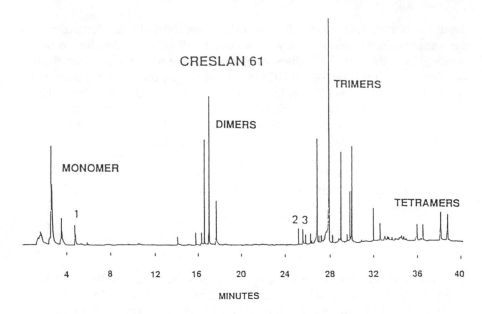

Figure 17.
Pyrogram of a polyacrylonitrile fiber (Creslan 61) from analysis by pyrolysis capillary gas chromatography. (From Challinor, J. M., in *Forensic Examination of Fibres*, Roberson, J., Ed., Horwood, London, 1992, 219. With permission.)

types, Nomex and Kevlar, could not be discriminated from each other. Py-MS had less discriminating power than IR for other types of fibers, however.

E.   Synthetic Polymers/Plastics

Synthetic polymeric materials occur frequently as evidential materials (bits of tape, rubber, plastic bags, automobile trim, and such) and, no doubt, constitute a significant portion of the samples being examined by Py-GC, Py-MS, and Py-GC/MS. There are published reports showing Py-GC comparisons of such materials, but this author did not find published examples of forensic analysis/comparisons of these sorts of materials by Py-MS or Py-GC/MS (other than those that have been mentioned in the other subsections) that would be appropriate for inclusion here. Py-MS and Py-GC/MS have been shown to be powerful tools for the characterization of synthetic polymeric materials, and any recent review of analytical pyrolysis will lead the interested reader to many examples.

F.   Human Hair

The microscopical comparison of human hair is a tedious, time-consuming process. In forensic laboratories where the number of hair com-

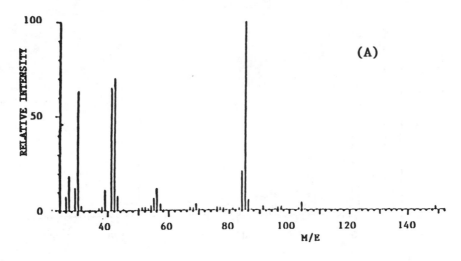

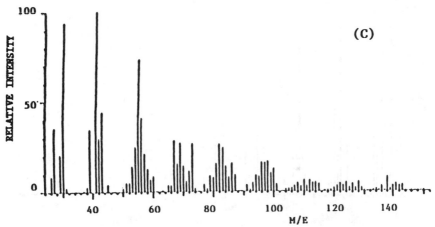

Figure 18.
Py-MS mass pyrograms of nylon fibers pyrolyzed at 600°C: (A) nylon 4; (B) nylon 6; (C) nylon 6.10; and (D) nylon 11. (From Hughes, J. C., Wheals, B. B., and Whitehouse, M. J., *Analyst*, 103, 482, 1978. With permission.)

parisons each year is small, few examiners are able to achieve the necessary skill level to perform satisfactory comparisons. This problem has led to a continual desire for instrumental methods for hair comparisons. Capillary Py-GC/MS and Py-GC/MS/MS were applied to this problem for several years at the FBI FSRG.[30,39] The early work was quite promising, with the Py-GC/MS profiles from different individuals showing striking differences. Figure 19 shows such a comparison between segments of unwashed hair from three donors. A considerable amount of effort was expended identifying various peaks in the profiles and speculating upon their origin in the hair.

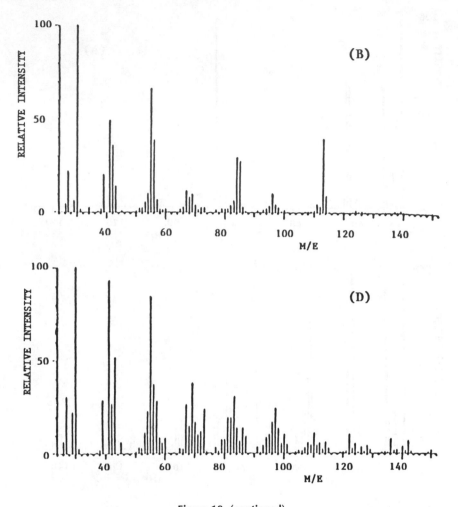

Figure 18. (continued)

Because it was anticipated that evidential hair would likely be contaminated with material which would need to be removed prior to analytical comparison to suspect hair, the final phase of this project used solvent-washed hair samples. A single hair from 12 individuals and 12 hairs from a single individual were examined. Most of the peaks in the pyrograms which had shown person-to-person differences in the unwashed hair studies were still present in the pyrograms of the solvent-washed hair. However, the pattern of these peaks was not found to be consistent in the 12 samples taken from the same individual. In fact, the same sort of variability seemed to occur within the set of samples from the same individual as occurred within the set of samples from the different indi-

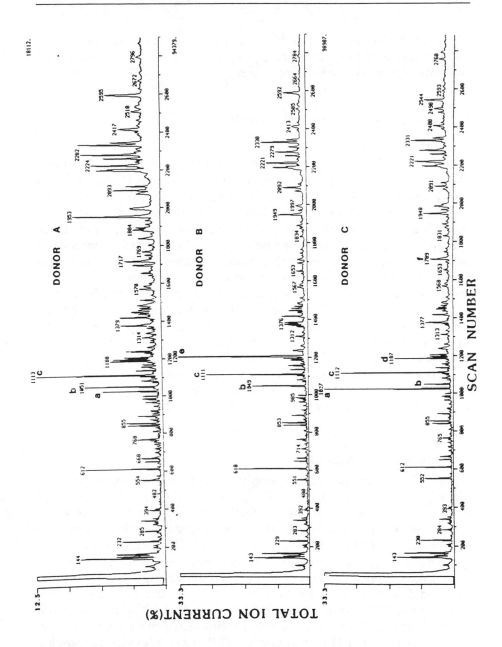

Figure 19.
Py-GC/MS pyrograms of hair from three donors (A, B, and C) showing differences among the three in the components labeled a through f. The pyrolyzates were separated on a DB-Wax capillary column held at 35°C for 0.5 min, heated to 180°C at 15°C/min, then to 250°C at 7°C/min. (From Munson, T. O. and Vick, J., *J. Anal. Appl. Pyrol.*, 8, 493, 1985. With permission.)

viduals.[40] It does not seem to this author that any of the studies to date have demonstrated that analytical pyrolysis, in any of its forms, is a useful technique for forensic hair comparisons.

G. Miscellaneous Samples and Techniques

1. Photocopy Toners

The Py-GC/MS analysis of toner material removed from 62 photocopies made on different machines demonstrated that Py-GC/MS is a useful analytical technique for such comparisons.[41] In each case, the sample consisted of toner material lifted from a single letter on the photocopy page. Most of the toners examined fell into two general groups: those containing a large amount of styrene, usually with smaller amounts of butyl methacrylate or other methacrylates, or some combination of these chemicals; and those toners consisting mainly of some type of epoxy resin. Figure 20 shows the Py-GC/MS comparison of toner material from three photocopies. The Py-GC/MS profile shown for Minolta EP 870 toner exhibits the pattern typical of the group containing epoxy resin. The profile for IBM 85 shows methyl methacrylate (1.1 min), styrene (1.9 min), and butyl methacrylate (2.4 min), but with the styrene, somewhat atypically, in a smaller amount than the methacrylate esters. The profile for IBM Copier II shows this toner to be even more atypical, being composed of methyl methacrylate and butyl methacrylate, with styrene completely absent.

2. Simultaneous Methylation

The FSL group has applied simultaneous pyrolysis methylation capillary gas chromatography (SPM-GC) and SPM-GC/MS to evidential materials.[42-46] It remains to be seen whether this technique achieves widespread use. The technique can be easily performed by any laboratory with capabilities for Py-GC or Py-GC/MS without additional equipment requirements. Figure 21 shows the profiles produced in the analysis of a commercial phenolic resin (tert-butylphenol-formaldehyde resin) by Py-GC and SPM-GC. Mass spectral data indicates the presence of tert-butylphenol (TBP) (peak 1), and methyl- and dimethyl-substituted TBP (peaks 2 and 3). Peaks 7, 8, and 9 in the SPM-GC profile are the methyl ethers of TBP, methyl-TBP, and dimethyl-TBP. For some evidential samples, this procedure could be a valuable additional comparison.

3. Comparison of Soots

Studies using Py-GC/MS to examine soots formed from the combustion of various alkylbenzenes[47] and various plastics[48] (which were recently

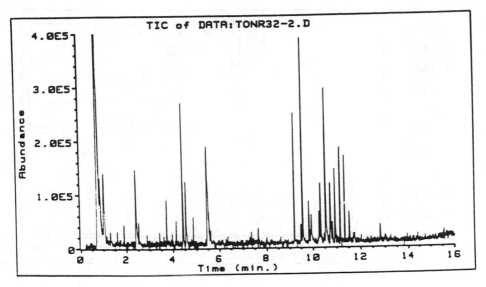

Class 16: Minolta EP 870

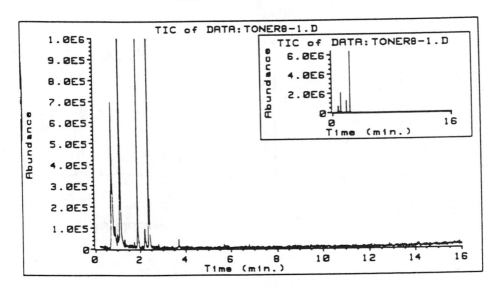

Class 17: IBM 85

Figure 20.
Py-GC/MS pyrograms from three photocopy toners. The total ion current profiles
are shown with the abundance scales magnified to show the smaller peaks, and with
all of the peaks on scale (inset). (From Munson, T. O., *J. Forensic Sci.*, 34, 352, 1989.
With permission.)

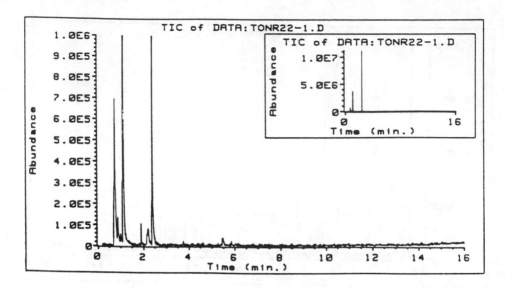

Class 18: IBM Copier II

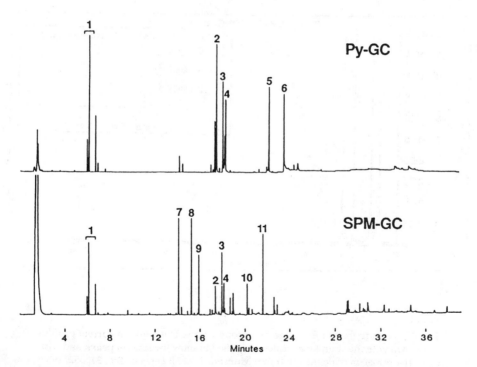

Figure 21.
Capillary Py-GC and simultaneous pyrolysis methylation capillary gas chromatography (SPM-GC) of TBP-formaldehyde resin. (From Challinor, J. M., *J. Anal. Appl. Pyrol.*, 25, 349, 1993. With permission.)

published by a forensic science laboratory in conjunction with a university in Japan) could possibly be useful to other forensic scientists. Although the text of the second work is written in Japanese, the inclusion of an English captioned figure presenting pyrograms from soots of 12 types of plastics, a 2-page table containing chromatographic peak identifications in English, and an English abstract may make this report quite useful to many persons not capable of reading Japanese. Figure 22 shows the Py-GC/MS profiles from the soots generated by combustion of four types of plastics: ABS, acrylonitrile/styrene/butadiene copolymer; UP, unsaturated polyester; PC, polycarbonate; and PPO, poly(phenylene oxide). As might be expected from soots, the bulk of the pyrolyzate (peaks 59 to 79) was made up of polynuclear aromatic hydrocarbons. In each case, however, there do appear to be components which are characteristic of the individual plastics.

V. CONCLUSIONS AND FUTURE DIRECTIONS

Clearly, forensic pyrolysis mass spectrometry has been shown to be a useful analytical technique for the examination and comparison of many materials of interest to the forensic scientist. Routine use of this technique for the examination of evidential materials, however, has been limited to very few laboratories in the world. In pondering why this technique has not become more widely used, one might suppose that a Py-GC/MS or Py-MS system might be too sophisticated and/or too expensive for routine use in a forensic science laboratory. This is clearly not the case; it was pointed out earlier that GC/MS systems were used in many forensic science laboratories in the mid-1970s.[12] In the U.S. today, no forensic science laboratory would be considered adequately equipped without a GC/MS system. However, these GC/MS systems are dedicated almost entirely to one task, drug analysis, and are not available for analysis of other materials.

One might wonder why forensic science laboratories acquire GC/MS systems for drug analysis but, only rarely, for forensic pyrolysis. This situation is not hard to understand when one considers the utility of the data from a GC/MS drug analysis compared to that of pyrolysis data. The GC/MS identification of a controlled substance can be a direct proof of guilt in a court proceedings. Although material comparisons and identifications by pyrolysis mass spectrometry can provide very useful investigative information, these data would not provide direct proof of guilt, but rather would add to the weight of evidence of guilt or innocence.

Py-GC is widely used, successfully, to characterize and compare evidential samples. The use of forensic pyrolysis mass spectrometry will not become more widespread until the technique provides better answers to

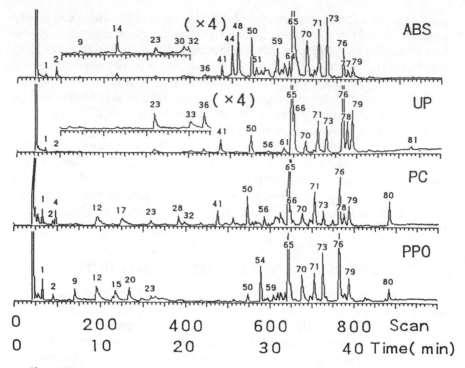

Figure 22.
Total ion current profiles obtained from the pyrolysis of four soots at 560°C followed by capillary GC/MS analysis. The plastics from which the soots were produced were ABS, acrylonitrile/styrene/butadiene copolymer; UP, unsaturated polyester; PC, polycarbonate; and PPO, poly(phenylene oxide). (From Takatsu, M. and Yamamoto, T., *Bunseki Kagaku*, 42, 543, 1993. With permission.)

the question concerning how significant it is to find two evidential samples analytically indistinguishable. Because mass spectrometry adds several dimensions of information not available in Py-GC data, there is considerable potential for development in this area. Improvements seem most likely to come in the areas of data storage, handling, and interpretation.

REFERENCES

1. Wheals, B. B., *J. Anal. Appl. Pyrol.*, 2, 277, 1981.
2. Wheals, B. B., in *Analytical Pyrolysis*, Jones, C. E. R. and Cramers, C. A., Eds., Elsevier, Amsterdam, 1977, 89.
3. Irwin, W. J. and Slack, J. A., *Analyst*, 103, 673, 1978.
4. Nelsen, D. G., Yee, J. L., and Kirk, P. L., *Microchem. J.*, 6, 225, 1962.
5. Zemany, P. D., *Anal. Chem.*, 25, 1709, 1953.

6. De Forest, P. R., D. Crim. thesis, University of California, Berkeley, CA, 1969.
7. De Forest, P. R., *J. Forensic Sci.*, 19, 113, 1974.
8. Martin, S. B., *J. Chromatogr.*, 2, 273, 1959.
9. Lehrle, R. S. and Robb, J. C., *Nature*, 183, 1671, 1959.
10. Radell, E. A. and Strutz, H. C., *Anal. Chem.*, 31, 1890, 1959.
11. Noble, W., Wheals, B. B., and Whitehouse, M. J., *Forensic Sci. Int.*, 3, 163, 1974.
12. Hughes, J. C., Wheals, B. B., and Whitehouse, M. J., *Forensic Sci. Int.*, 10, 217, 1977.
13. Meuzelaar, H. L. C. and Kistemaker, P. G., *Anal. Chem.*, 45, 587, 1973.
14. Meuzelaar, H. L. C., Posthumus, M. A., Kistemaker, P. G., and Kistemaker, J., *Anal. Chem.*, 45, 1546, 1973.
15. Meuzelaar, H. L. C., Kistemaker, P. G., and Posthumus, M. A., *Biomed. Mass Spectrom.*, 1, 312, 1974.
16. Saferstein, R. and Manura, J., *J. Forensic Sci.*, 22, 748, 1977.
17. Hughes, J. C., Wheals, B. B., and Whitehouse, M. J., *Analyst*, 102, 143, 1977.
18. Hughes, J. C., Wheals, B. B., and Whitehouse, M. J., *Analyst*, 103, 482, 1978.
19. Hickman, D. A. and Jane, I., *Analyst*, 104, 334, 1979.
20. Ardrey, R. E., Batchelor, T. M., and Smalldon, K. W., *Central Research Establishment Report No 311*, 1979, Aldermaston, England.
21. Ardrey, R. E., Batchelor, T. M., and Smalldon, K. W., *Central Research Establishment Report No 322*, 1979, Aldermaston, England.
22. Irwin, W. J., *J. Anal. Appl. Pyrol.*, 1, 89, 1979.
23. Wheals, B. B., *J. Anal. Appl. Pyrol.*, 2, 277, 1981.
24. Brettell, T. A. and Saferstein, R., *Anal. Chem.*, 55, 19R, 1983.
25. Saferstein, R., in *Pyrolysis and GC in Polymer Analysis*, Liebman, S. A. and Levy, E. J., Eds., Marcel Dekker, New York, 1985, 339.
26. Castle, D. A., Curry, C. J., and Russell, L. W., *Forensic Sci. Int.*, 24, 285, 1984.
27. Burke, P., Curry, C. J., Davies, L. M., and Cousins, D. R., *Forensic Sci. Int.*, 28, 201, 1985.
28. Bakowski, N. L., Bender, E. C., and Munson, T. O., *J. Anal. Appl. Pyrol.*, 8, 483, 1985.
29. McMinn, D. G., Carlson, T. L., and Munson, T. O., *J. Forensic Sci.*, 30, 1064, 1985.
30. Munson, T. O. and Vick, J., *J. Anal. Appl. Pyrol.*, 8, 493, 1985.
31. Whitehouse, M. J., Boon, J. J., Bracewell, J. M., Gutteridge, C. S., Pidduck, A. J., and Puckey, D. J., *J. Anal. Appl. Pyrol.*, 8, 515, 1985.
32. Wheals, B. B., *J. Anal. Appl. Pyrol.*, 8, 503, 1985.
33. Levy, E. J. and Wampler, T. P., in *Proceedings of the International Symposium on Questioned Documents*, U.S. Government Printing Office, Washington, D.C., 1985, 141.
34. Levy, E. J. and Wampler, T. P., *J. Forensic Sci.*, 31, 258, 1986.
35. Munson, T. O., *Crime Lab. Digest*, 13, 82, 1986.
36. Curry, C. J., *J. Anal. Appl. Pyrol.*, 11, 213, 1987.
37. Wright, M. M. and Wheals, B. B., *J. Anal. Appl. Pyrol.*, 11, 195, 1987.
38. Munson, T. O., *Crime Lab. Digest*, 14, 112, 1987.
39. Munson, T. O. and Fetterolf, D. D., *J. Anal. Appl. Pyrol.*, 11, 15, 1987.
40. Munson, T. O., *Crime Lab. Digest*, 14, 153, 1987.
41. Munson, T. O., *J. Forensic Sci.*, 34, 352, 1989.
42. Challinor, J. M., *J. Anal. Appl. Pyrol.*, 16, 323, 1989.
43. Challinor, J. M., *J. Anal. Appl. Pyrol.*, 18, 233, 1991.
44. Challinor, J. M., *J. Anal. Appl. Pyrol.*, 20, 15, 1991.
45. Challinor, J. M., in *Forensic Examination of Fibres*, Roberson, J., Ed., Horwood, London, 1992, 219.
46. Challinor, J. M., *J. Anal. Appl. Pyrol.*, 25, 349, 1993.
47. Takatsu, M. and Yamamoto, T., *Anal. Sci.*, 9, 267, 1993.
48. Takatsu, M. and Yamamoto, T., *Bunseki Kagaku*, 42, 543, 1993.

49. Chiavari, G., Galletti, G. C., Lanterna, G., and Mazzeo, R., *J. Anal. Appl. Pyrol.*, 24, 227, 1993.
50. Shedrinsky, A. M., Grimaldi, D. A., Boon, J. J., and Baer, N. S., *J. Anal. Appl. Pyrol.*, 25, 77, 1993.
51. Grob, K., Jr., Grob, G., and Grob, K., *J. Chromatogr.*, 156, 1, 1978.
52. Meuzelaar, H. L. C., Windig, W., Harper, A. M., Huff, S. M., McClennen, W. H., and Richards, J. M., *Science*, 226, 268, 1984.
53. Wheals, B. B., in *Proceedings of the Int. Symp. on the Analysis and Identification of Polymers*, Federal Bureau of Investigation, Quantico, VA, 1985.
54. Munson, T. O., Carlson, T. L., and McMinn, D. G., in *Proceedings of the Int. Symp. on the Analysis and Identification of Polymers*, Federal Bureau of Investigation, Quantico, VA, 1985.

Chapter 6

Detection and Identification of Explosives by Mass Spectrometry

Dean D. Fetterolf

CONTENTS

Note: This is Publication 93-14 of the Laboratory Division of the Federal Bureau of Investigation. Names of commercial manufacturers are provided for identification only and do not constitute or imply endorsement, recommendation or favoring by the Federal Bureau of Investigation.

I. INTRODUCTION TO EXPLOSIVES

The detection and identification of explosives by mass spectrometry involves a wide array of interesting and challenging analytical problems. Mass spectrometry is only one of many analytical tools used for the detailed characterization of commercial or military explosives and propellants. The forensic identification of explosives by mass spectrometry has been the subject of a recent review,[1] chapter,[2] and a new book by Yinon and Zitrin.[3]

The screening of evidentiary material for explosive residues, more properly called trace physical evidence, may provide an important investigative link between a suspect and the crime scene. A very thorough review of the analytical methods for the detection and identification of explosive residue has been published by Beveridge.[4] The detection of hidden explosives in aviation security presents a different set of analytical challenges. While either of the above scenarios presents a different analytical challenge, there are a number of fundamental similarities that must be thoroughly understood. These fundamentals include the types of explosives as well as their basic chemical and physical properties.

A. Types of Explosives

An explosive is a material which undergoes a very rapid self-propagating decomposition or chemical reaction resulting in the formation of intense heat, and the development of a sudden pressure or shock wave through the liberated and surrounding air.

Explosives of interest to the forensic mass spectroscopist can be divided into three main classes: high explosives, propellants ("low explosives"), and primary explosives. The first two are of most interest.

The organic high explosives are further subdivided into three groups:

1. The aromatic nitro compounds characterized by $C-NO_2$ groups, e.g., trinitrotoluene (TNT) and tetryl
2. The nitramines characterized by $N-NO_2$ groups, e.g., cyclotrimethylene trinitramine (RDX) and cyclotetramethylene tetranitramine (HMX)
3. The nitrated esters characterized by $C-ONO_2$ groups, nitroglycerin (NG), ethylene glycol dinitrate (EGDN), and pentaerythritol tetranitrate (PETN)

High explosives also include inorganic-based explosives such as ammonium and nitrate, gels, and emulsion explosives which are most commonly used in commercial mining and construction. While these explosives are of great forensic importance, their characterization and residue analysis is most often done with liquid chromatographic methods. An extensive review of these LC method has been published by Lloyd.[5]

Figure 1 shows the chemical structures of the common organic explosives. These explosives, when initiated, detonate. Detonation is a chemical reaction which produces a shock wave with high temperature and pressure gradients being generated. These explosives are rarely encountered

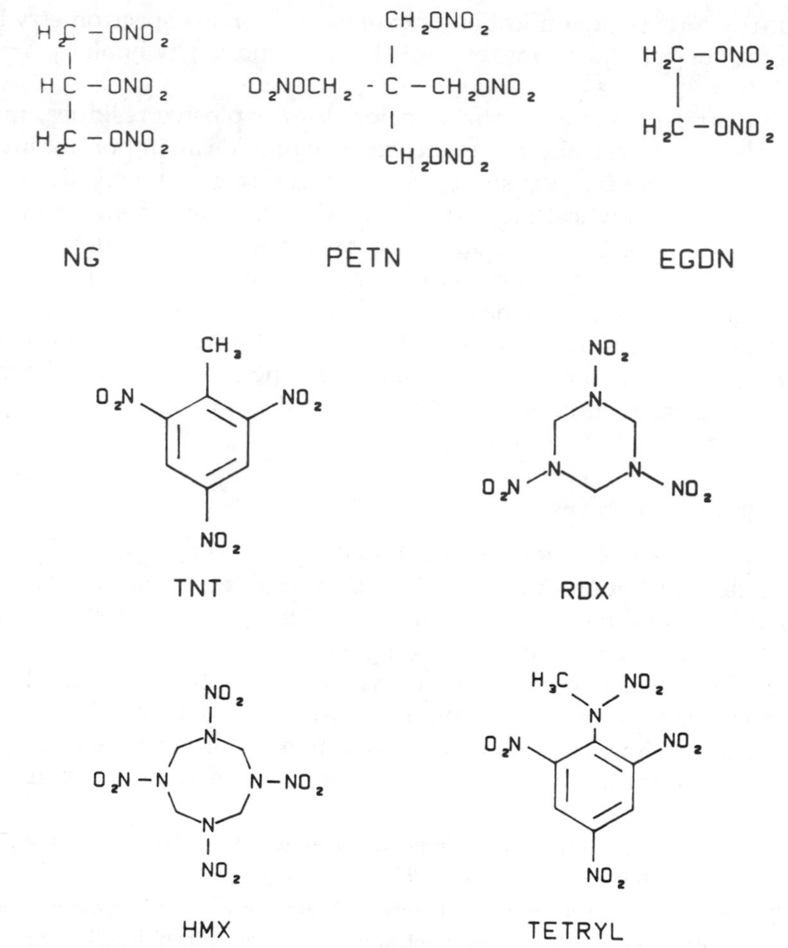

Figure 1.
Chemical structure of common explosives.

in their pure form. C-4, a common U.S. military explosive, is a mixture of RDX, plasticizers, and oil. Nitrated dynamites often contain mixtures of NG and EGDN. Mixtures of PETN and RDX are most often associated with a Czechoslovakian explosive, SEMTEX.

Propellants are generally associated with ammunition. Three types of propellants exist based upon their composition. Single-based powders consist primarily of nitrocellulose. Double-based powders consist of nitrocellulose and nitroglycerin. Triple-based powders are double-based powders with nitroguanidine added. Nitroguanidine acts as a flash suppressor. Stabilizers, antioxidants, and static control chemicals are also used. These components aid in the differentiation of brands of smokeless powder as will be discussed later. Low explosives generally deflagrate or burn in unconfined spaces.

Primary explosives such as lead styphnate or mercury fulminate detonate by simple ignition from a spark, flame, or impact. Primary explosives are most often associated with the primers in ammunition and blasting caps. The primary explosives are not amenable to organic mass spectrometry analysis.

The well-informed forensic mass spectroscopist is also familiar with the increasing number of publications dealing with homemade explosives and incendiary devices. Often neglected by detector manufacturers and even some agencies, the criminal use of these chemicals can produce equally as violent, deadly, and devastating results.

B. Effects of Chemical and Physical Properties of Explosives on Detection

Detailed chemical and physical properties and manufacturing methods of explosives are best gleaned from the four-volume set by Urbanski entitled *Chemistry and Technology of Explosives*[6] or the reference text by Meyer entitled *Explosives.*[7]

Conrad[8] has identified four fundamental properties that are shared by the common organic explosives. Electronegativity, adsorbtivity, thermal stability, and frangibility of explosives greatly affect their behavior in terms of sampling and analysis by any analytical technique including mass spectrometry. A fifth property, vapor pressure, is also crucial. The widely varying and extremely low vapor pressure cannot be ignored. The uninitiated and, at times, the experienced analyst or explosive detector designer often goes astray by failing to recognize these common traits.

1. Electronegativity

The common explosives with their nitro and nitramine groups exhibit a strong electron affinity. This makes them excellent candidates for electron capture or other negative ionization methods. From an explosive detector standpoint, this property increases the selectivity of the detector because many environmental contaminants such as hydrocarbons do not readily produce negative ions.

2. Adsorbtivity

The highly polar explosive molecules tenaciously adsorb to many surfaces such as steel, wood, glass quartz, fused silica, and even teflon.[9] Once adsorbed the explosives must be liberated in order to be sampled and analyzed. To prevent unwanted adsorption, temperatures as high as

140°C must be maintained, but, at temperatures greater than this, one risks thermal decomposition. As discussed in Section III, the forensic mass spectroscopist can take advantage of this, both as a means of trapping explosive vapors or as an aid in locating hidden explosives.

3. Thermal Stability

Explosives by their very nature are thermally unstable. While heating a surface or adsorbent is a common means of liberating trapped material, it must be done cautiously to prevent degradation of the analyte. High temperatures in the injector of a gas chromatograph or ion source in the mass spectrometer are likely places for thermal degradation of explosives. Mass spectrometry has recently been used to study the thermal degradation of RDX and HMX.[10]

4. Frangibility

The addition of small amounts of energy to explosives will rupture chemical bonds. This is most evident in the electron impact mass spectra which exhibit no molecular ions. Extensive fragmentation often occurs with chemical ionization as well. Extensive fragmentation of TNT and tetryl was also observed with 5-ns laser irradiation at 266 nm followed by time of flight mass analysis.[11]

5. Vapor Pressure

Table 1 shows the equilibrium vapor pressures of the common explosives at standard conditions of temperature and pressure. A detailed discussion of the measurement of the vapor pressure of explosives has been published.[12] It is interesting to note the nearly eight orders of mag-

TABLE 1.

Equilibrium Vapor Pressures of
Common Explosives at 25°C

Explosive	Vapor pressure
NG	580 ppb
NH_4NO_3	12 ppb
TNT	9.4 ppb
PETN	18 ppt
RDX	6 ppt

Adapted from Dionne, B. C., Rounbehler,
D. P., Achter, E. K., Hobbs, J. R., and
Fine, D. H., *J. Energ. Materials*, 4, 447,
1986.

nitude difference in the vapor pressures of EGDN and HMX. Vapor pressure alone represents a general indication of the retention order in most gas chromatographic separations.

The implications of this widely varying vapor pressure and its role in sampling and detector design cannot be overlooked. High vapor pressure explosives may quickly saturate a system designed for ultimate sensitivity. For the low vapor pressure explosives little vapor may be present.

C. Conventional Ionization of Explosives

The conventional ionization of explosives by electron impact and positive and negative chemical ionization has been thoroughly reviewed by Yinon.[13] As described above, the frangibility of explosives is demonstrated in the extensive fragmentation observed in electron impact. Chemical ionization mass spectra are strongly dependent upon source temperature, reagent gas identity, and pressure. Therefore, it is essential that standards must be run on each individual mass spectrometer system at the time samples are analyzed. To demonstrate these points several EI and CI mass spectra are discussed below.

1. Electron Impact

The electron impact mass spectra of NG, TNT, PETN, and RDX are shown in Figure 2. The nitrate esters NG, PETN, and EGDN (not shown) are characterized by only two ions m/z 46 $(NO_2)^+$ and m/z 62 $(NO_3)^+$. The molecular ion is never observed. TNT exhibits a strong loss of OH radical to produce m/z 210. RDX, one of the most common military explosives, exhibits dominant ions at m/z 42 $(CH_2NCH_2)^+$ and m/z 46 $(NO_2)^+$. In the midmass region, m/z 148 $(M–CH_2NNO_2)^+$ is formed by ring contraction followed by subsequent loss of CH_2N to form m/z 120.

The extensive fragmentation of explosives was also observed in the negative ion mode for RDX, PETN, and TNT using a reversal electron attachment device technique (READ).[14] In the READ technique, electrons are brought to near zero energy by reversing their direction with electrostatic fields. Under single-collision conditions electron capture then takes place. The ions are focused and deflected by a 90° electrostatic analyzer (ESA) into a quadrupole mass spectrometer. Unfortunately, the time required to record the spectra in this instrument ranged from 0.5 to 2 h, thus limiting its applicability at this time.

2. Chemical Ionization

The positive and negative chemical ionization mass spectra of NG, TNT, PETN, and RDX as recorded on a Finnigan MAT TSQ 700 triple stage

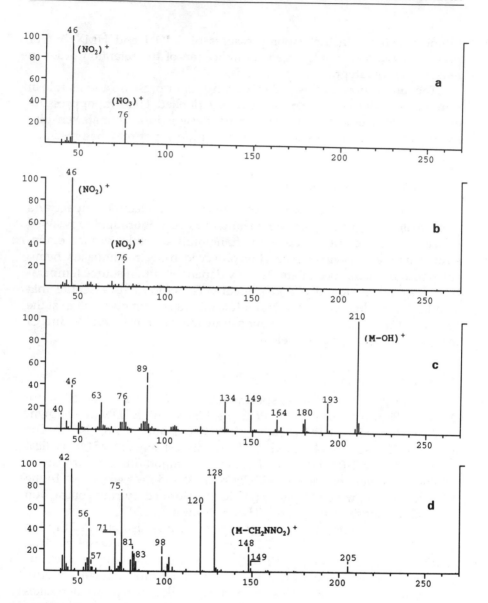

Figure 2.
Electron impact mass spectra of (a) NG, (b) PETN, (c) TNT, and (d) RDX.

quadrupole mass spectrometer at a source temperature of 120°C and methane pressure of 0.8 torr are shown in Figures 3 and 4. The high electronegativity of the explosives makes negative ion chemical ionization the method of choice when maximum sensitivity is required.

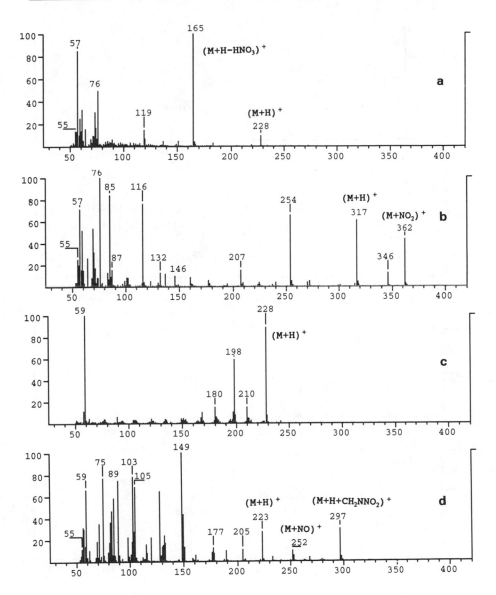

Figure 3.
Positive ion methane chemical ionization of (a) NG, (b) PETN, (c) TNT, and (d) RDX.

In the positive ion mode, nitroglycerin exhibits a weak protonated molecular ion at m/z 228 and a base peak at m/z 165 formed by the subsequent loss of nitric acid. The negative ion spectra is characterized by a base peak of m/z 62 (ONO_2)$^-$ and a weak (M + 62)$^-$ at m/z 289.

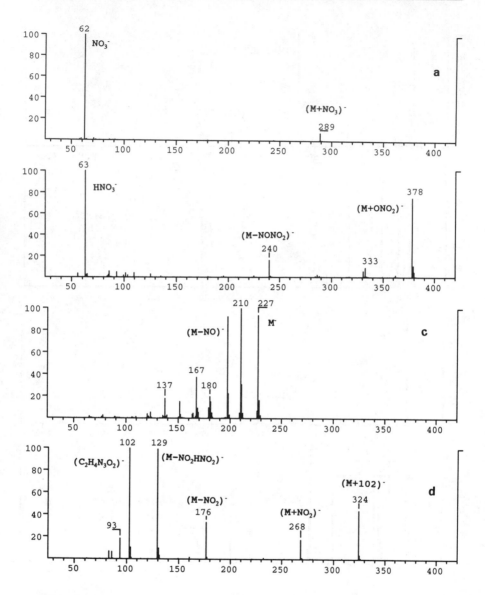

Figure 4.
Negative ion methane chemical ionization of (a) NG, (b) PETN, (c) TNT, and (d) RDX.

The positive ion methane CI spectrum of PETN exhibits a number of adduct ions, namely the protonated molecular ion at m/z 317, and (M + NO)$^+$ at m/z 346 and (M + NO$_2$)$^+$ at m/z 362. The loss of HONO$_2$ from (M + H)$^+$ forms m/z 254. A host of fragment ions exists at the lower masses. In the negative ion mode, a base peak of m/z 63 corresponds to nitric acid $(HNO_3)^-$. The fragment at m/z 240 is $(M–NONO_2)^-$. No molecular anion is observed; however, oxygen adducts at m/z 331 and m/z 333 correspond to the $(M–H + O)^-$ and $(M + H + O)^-$, respectively. The ion at m/z 378 is $(M + ONO_2)^-$.

TNT is characterized by a protonated molecular ion $(M + H)^+$ at m/z 228 followed by the loss of water (m/z 210) and reduction to form m/z 198. In the negative ion mode, electron capture dominates forming m/z 227 followed by loss of OH to m/z 210. The loss of NO from the molecular anion forms m/z 197.

In the positive CI mass spectrum of RDX, a 30% relative abundance of $(M + H)^+$ at m/z 223 is observed. Adduct ions are formed between the molecular ion and NO to form m/z 252 and the protonated molecular ion and CH_2NNO_2 to form m/z 297. The base peak of m/z 149 is formed by ring opening and the loss of CH_2NNO_2 from the protonated molecular ion. A host of lower molecular weight fragment ions also exist. The negative ion spectra is characterized by adducts at m/z 268 $(M + NO_2)^-$ and m/z 324 $(M + 102)^-$. The fragment ion at m/z 176 corresponds to the loss of NO_2 while m/z 129 corresponds to the loss of NO_2HNO_2 from the molecular anion. The ion at m/z 102 is $(C_2H_4N_3O_2)^-$.

II. EXPLOSIVE CHARACTERIZATION

A. Gas Chromatography/Mass Spectrometry Analysis

The separation power of capillary gas chromatography and identification ability of mass spectrometry make it an ideal tool for explosive analysis. The use of gas chromatography/mass spectrometry (GC/MS) for the identification of two unusual explosives, postblast residue, headspace vapors, and hydrocarbon analysis in ammonium nitrate fuel oil mixtures has been published by Reutter and co-workers.[15] Zitrin has described the GC/MS confirmation of positive TLC screens in postblast analysis of PETN residue on a safe and RDX residue on a trash can.[16] Postblast NG and ethylcentralite were indicative of a smokeless powder-based improvised explosive device.[17] Problems were noted with the failure of PETN and RDX to elute in a reproducible way from the same 30-m capillary GC column.[17] The problems of thermal degradation of the nitrate esters NG and PETN in chromatographic systems were overcome by the use of a very short 1.5-m DB-5 column.[18] Hobbs has described the analysis of the explosive content, oils, dyes, plasticizers, and antioxidants as well as the EGDN and NG contamination of SEMTEX.[19]

1. Characterization of Trace Components in Explosive Mixtures

The separation power of capillary GC and identification ability of mass spectrometry make it an ideal tool for characterizing seized explosives in an attempt to determine a source of common origin. The organic extract contains the explosives, dyes, plasticizers, and other co-extracted additives.

a. TNT

Trace mono- and di-nitrotoluenes in TNT can be diagnostic in attempts to characterize these samples. The composition and mass spectra of impurities in military TNT vapors has been published.[20] It is interesting to note that while 2,4-DNT represents only 0.08% of the solid phase composition of TNT, the vapor phase composition is 35%, 2,4-DNT due to the dramatic differences in equilibrium vapor pressure.[20] A comparison of GC techniques and GC/MS for the identification of TNT and related nitroaromatics compounds has been published.[21] An HPLC procedure using photodiode array detection and cluster analysis has also been published by McCord.[22]

Trace analysis by GC can often be hampered by the chemical noise associated with column bleed or co-eluting analytes. Tandem mass spectrometry (MS/MS)[23,24] can greatly improve the limits of detection. The sample preparation procedure for GC/MS/MS involves the preferential extraction of the DNT isomers from 100 mg of TNT into boiling hexane. The hexane is cooled and decanted. GC conditions involve a 15-m × 0.25-mm DB Wax column temperature programmed from 150°C (1 min) to 250°C at 5°C/min with a 4-min hold time at 250°C.

Figure 5 shows the advantage of MS/MS in the reduction of chemical noise over GC/MS. In this case the protonated molecular ion m/z 183 was chosen. For example, the 2,6 isomer is barely visible in the reconstructed ion current (RIC) in the GC/MS trace, but is clearly well above background in the GC/MS/MS RIC. The daughter spectra of the isomers were easily distinguishable.

b. Plastic Bonded Explosives

The detailed physical examination of timing mechanisms, batteries, initiators, and other physical properties of the recovered terrorist devices by trained forensic explosives experts can lead to the development of a bomber's "signature". The forensic examination of terrorist devices was the subject of an international symposium.[25] Personal accounts of the criminal activity of a terrorist also describe the "signature" aspect of a bomber.[26]

The detailed chemical analysis of trace materials from the recovered explosives and a suspect source can also aid the forensic examiner in

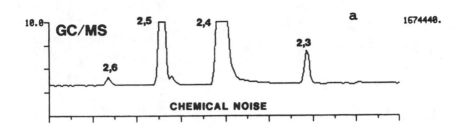

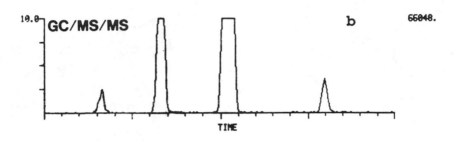

Figure 5.
Reconstructed ion chromatogram of (a) GC/MS and reduction in chemical noise by (b) GC/MS/MS.

determining a source of common origin. Extracts from samples of home-made PETN sheet explosive were analyzed by GC/MS; 10 mg of the explosive from the center of the sample were extracted with 1 ml of hexane.

A 15-m, 0.1-micron DB-5, 0.25-mm i.d. capillary column was tempera-ture programmed from 50°C (1 min hold) to 250°C at 10°C/min and held there for 4 min. A cold on-column injector was used. The Finnigan TSQ-45 triple stage quadrupole mass spectrometer was operated in the Q3 MS mode from 45 to 450 amu/s with an electron multiplier voltage of 1200 V. An ion source temperature of 150°C and transfer line temperature of 200°C were used.

Four major compounds were observed at significant total ion current count rates. Three of these are plasticizers and are common additives in the manufacturing of plastic and rubber. The structure of these three phthlates is shown in Figure 6. They are easily recognizable by the pres-ence of the m/z 149 ion, thus making it easy to identify them, based upon extracted ion current profiles. The fourth compound, also identified by library searching, was 1,3-benzenediol, monobenzoate. A sample of DuPont FLEX-X, a PETN-based sheet explosive, was found to contain 1,2,3-propanetricarboxylic acid, 2-(acetyloxy)-, tributyl ester, a patented addi-tive used in the manufacturing of FLEX-X.

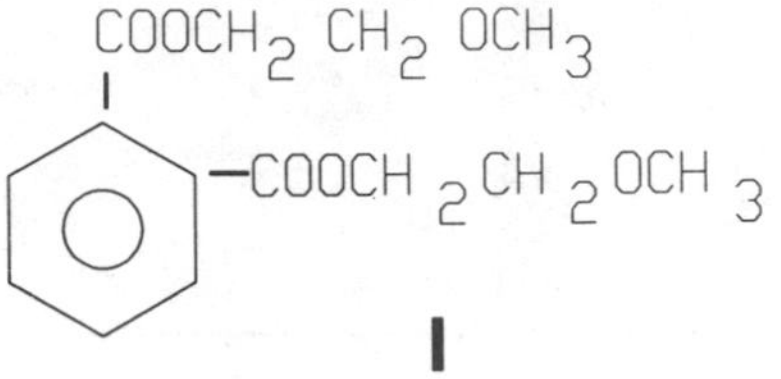

I

1,2-Benzenedicarboxilic Acid,
Bis(2-Methoxyethyl) Ester (BMP)

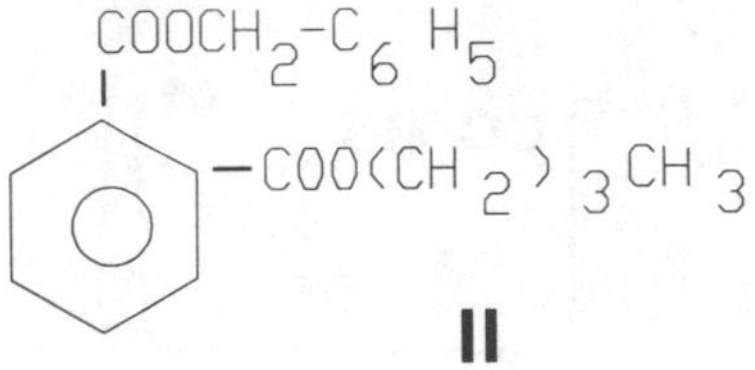

II

1,2-Benzenedicarboxylic Acid,
Butyl Phenylmethyl Ester (BPP)

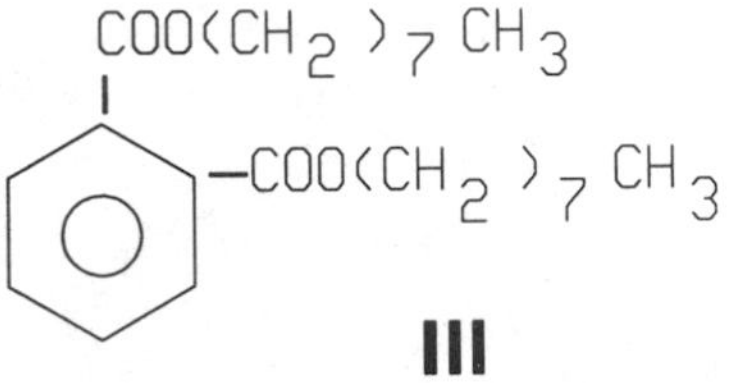

III

1,2-Benzenedicarboxylic Acid,
Diisootyl Ester (DIP)

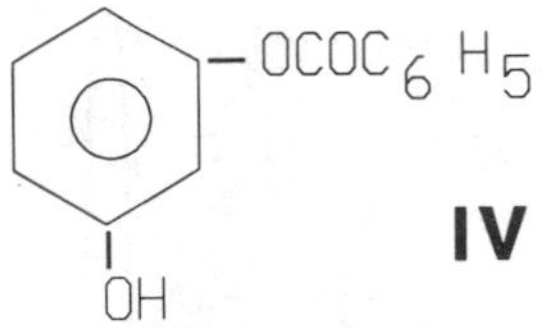

IV

1,3-Benzenediol,
Monobenzoate

Figure 6.
Chemical structure of compounds identified in homemade PETN sheet explosive.

Figure 7 shows the RICs of one group of three samples which was characterized by the presence of the three phthlates (I to III) and compound IV. Investigative data associated with the recovery of these explosives indicated that they may have had a common origin. Such analysis could lead to the development of a database of RICs of the chemical content of seized explosives. Explosives recovered in subsequent investigations can then be compared in attempts to aid the investigator in determining the origin of the explosive material.

c. *Smokeless Powders*

The detailed physical and chemical characterization of smokeless powders is an important area in forensic explosives analysis. Smokeless powders are most often associated with ammunition. As reloading supplies, they are widely available in gun stores and often find criminal uses beyond their intended purpose. When properly confined in an improvised explosive device (IED), smokeless powders can lead to devastating results. Smokeless and black powder IEDs make up more than 58% of the criminal bombings in the U.S.[27] Their detailed characterization involves measurements of particle morphology, shape, size, and chemical composition and represent a significant investigative tool in pipe bombings.[28]

A simple method for the organic analysis of smokeless powders by GC/MS has been developed by Martz and co-workers.[29] A few particles are extracted in 0.5 ml chloroform for 10 min with vortexing; 1 μl of solvent

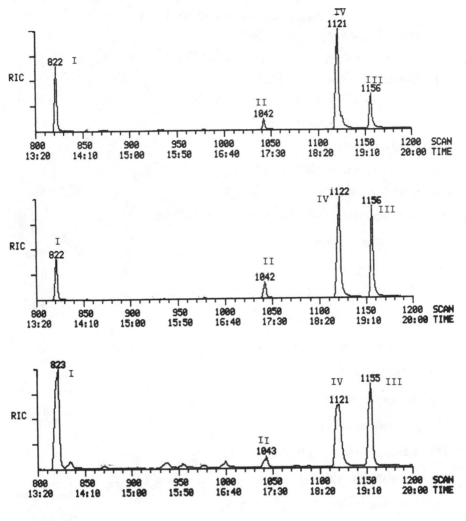

Figure 7.
Comparison of RICs of three seized homemade PETN sheet explosives.

is analyzed using a cold on-column injection on a 20-m SE-54 or DB-5 column.

Figure 8 shows the range of compounds possible in smokeless powder extracts. They consist of plasticizers (phthalates) and stabilizers (diphenyl amine, methyl, and ethylcentralite). The nitro and nitroso derivatives of diphenylamine are formed by its reaction with the degrading nitrocellulose.

Figure 9 shows the analysis of 4 different smokeless powders from a library of over 100. The comparison of chromatographic profiles from so many samples would be a time-consuming procedure. Martz and co-

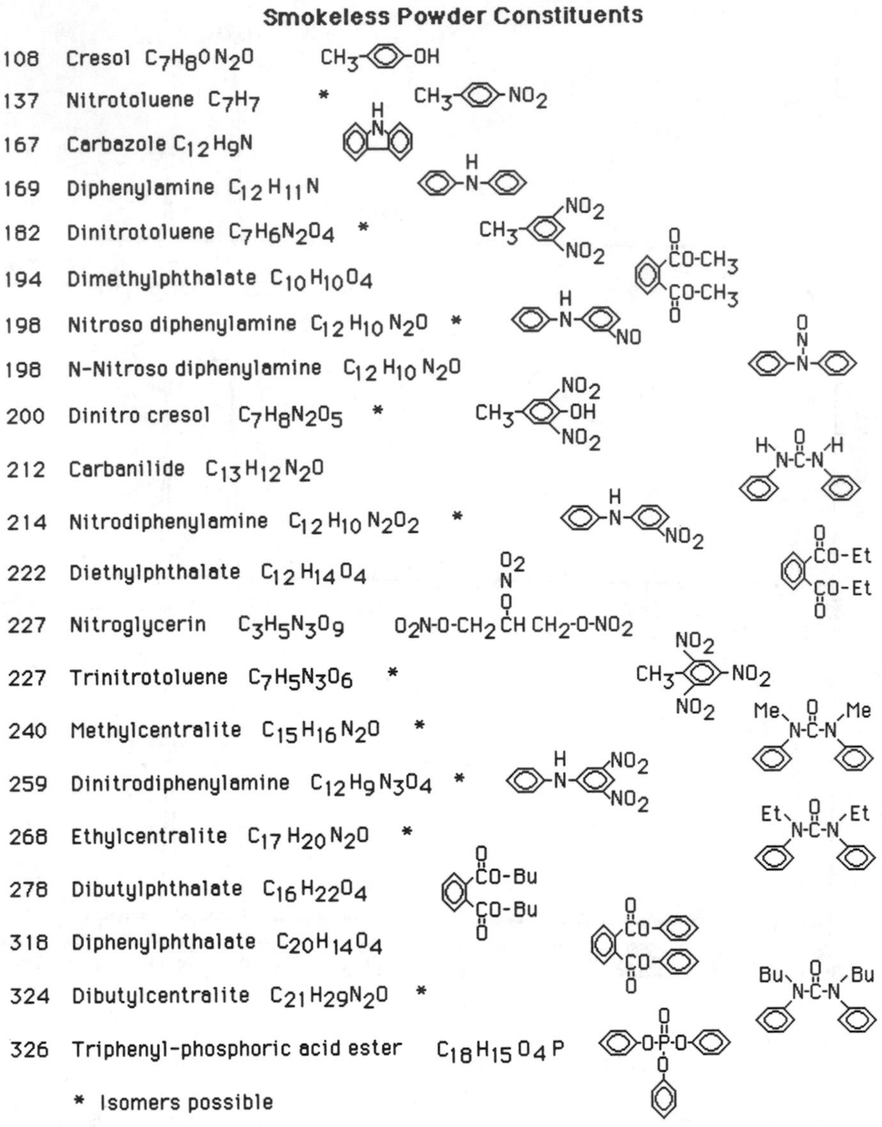

Smokeless Powder Constituents

108　Cresol $C_7H_8ON_2O$　　　　CH_3-⬡-OH

137　Nitrotoluene C_7H_7　　*　　CH_3-⬡-NO_2

167　Carbazole $C_{12}H_9N$

169　Diphenylamine $C_{12}H_{11}N$　　⬡-N-⬡

182　Dinitrotoluene $C_7H_6N_2O_4$　*　　CH_3-⬡$\genfrac{}{}{0pt}{}{NO_2}{NO_2}$

194　Dimethylphthalate $C_{10}H_{10}O_4$

198　Nitroso diphenylamine $C_{12}H_{10}N_2O$　*　⬡-N-⬡-NO

198　N-Nitroso diphenylamine $C_{12}H_{10}N_2O$

200　Dinitro cresol $C_7H_8N_2O_5$　*　CH_3-⬡$\genfrac{}{}{0pt}{}{OH}{NO_2}$ (with NO_2 top)

212　Carbanilide $C_{13}H_{12}N_2O$

214　Nitrodiphenylamine $C_{12}H_{10}N_2O_2$　*　⬡-N-⬡-NO_2

222　Diethylphthalate $C_{12}H_{14}O_4$

227　Nitroglycerin $C_3H_5N_3O_9$　　O_2N-O-CH_2 CH CH_2-O-NO_2

227　Trinitrotoluene $C_7H_5N_3O_6$　*

240　Methylcentralite $C_{15}H_{16}N_2O$　*

259　Dinitrodiphenylamine $C_{12}H_9N_3O_4$　*

268　Ethylcentralite $C_{17}H_{20}N_2O$　*

278　Dibutylphthalate $C_{16}H_{22}O_4$

318　Diphenylphthalate $C_{20}H_{14}O_4$

324　Dibutylcentralite $C_{21}H_{29}N_2O$　*

326　Triphenyl-phosphoric acid ester　$C_{18}H_{15}O_4P$

* Isomers possible

Figure 8.
Chemical structure of compounds identified in smokeless powder extracts.

workers have developed a procedure in which the reconstructed ion current profile is processed by summing all the mass spectra and storing the composite spectrum in a mass spectral library of similarly analyzed smokeless powder samples. Using the Finnigan INCOS library search software, unknowns can then be compared. Interestingly, physical measurements of shapes and sizes were stored as searchable parameters usu-

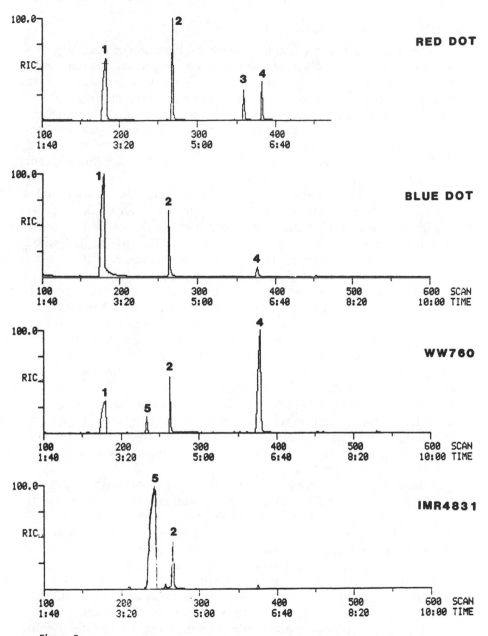

Figure 9.
Comparison of four double-based smokeless powders. Peak (1) nitroglycerin, (2) diphenylamine, (3) ethylcentralite, (4) dibutylphthalate, and (5) dinitrotoluene.

ally associated with molecular formula or weight. Limiting the mass spectral search to those powders with the measured physical properties greatly enhanced the search result.

d.   Organic Gunshot Residue

The most common method to indicate the handling or discharge of a firearm involves the analysis of hand swabs for antimony, barium, and/ or lead, components of the primer mixture.[30] Another popular method involves particle analysis of primer residue by scanning electron microscopy.[31] Interpretation of the forensic significance of the elemental concentrations is compounded by their ubiquitous environmental presence. Particle analysis methods are time consuming. The feasibility of organic gunshot residue detection was demonstrated by Mach.[32] More recent procedures involve the use of HPLC.[33]

As described above, the majority of firearms propellants are double-based in composition: containing nitroglycerin (NG), nitro cellulose, and stabilizers such as ethylcentralite or diphenylamine. Typical double-based ammunition ranges from 10 to 40% by weight NG. Due to its limited usage, NG is not a common environmental contaminant. Using a solid phase extraction and GC/MS, a rapid and simple procedure for the detection of NG on hand swabs was developed.[34] Cotton swabs moistened with methyl tertiary butyl ether (MTBE) are rubbed across the back of the hands of suspected shooters. NG is extracted from the swabs with 2- × 1-ml pentane washes. Sample cleanup is carried out using a preconditioned (pentane/ethyl acetate) disposable diol (Supelclean, LC-Diol) extraction column. Following a 1-ml pentane wash, the NG is eluted with 1 ml ethylacetate into a silanized screw cap vial.

The diol columns exhibited complete recovery of the NG standard. The effect of contamination was studied by spiking swabs with 92 ng of NG along with 100 µl of contaminants. Recoveries from swabs contaminated with gasoline, hand lotion, and floor sweeping extracts were 80, 85, and 87%, respectively.

Using cold on-column injection, 1 µl of the eluant was chromatographed on a 0.1-µm film, 15-m DB-5 × 0.25-mm i.d. capillary column temperature programmed from 100 to 250°C. Multiple ions were monitored at m/z 46 $(NO_2)^-$ and m/z 62 $(NO_3)^-$ using electron capture negative chemical ionization (methane). A comparison of EI (m/z 46, 76), positive chemical ionization (m/z 46, 76), and negative chemical ionization (m/z 46, 62) showed a relative sensitivity of 1.0, 1.6, and 352.

The Finnigan-MAT TSQ-45 mass spectrometer operated in the Q3 MS mode showed detection limits of 0.3 pg NG. Considering dilution and sample size, this equates to 0.3 ng of NG on the hand of a shooter. Figure 10 shows a comparison of a handblank, three shots with an S&W .38 special caliber revolver, and a 92 pg NG standard.

The highest amounts of NG are typically recovered from the thumb-web-forefinger area of the shooting hand. Test-firing a variety of weapons showed deposits ranging from 1.2 to 1600 ng. A number of issues related

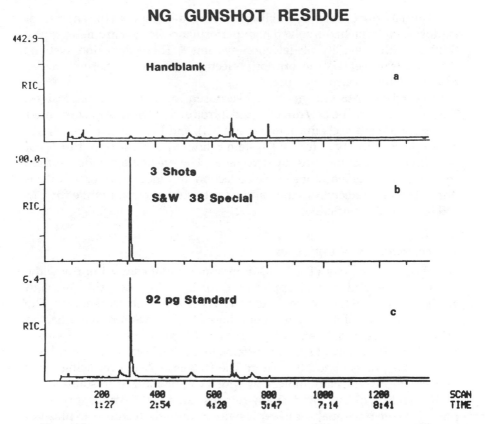

Figure 10.
Detection of NG on (a) handblank, (b) 3 shots with an S&W 38 special, and (c) 92 pg
NG standard.

to the forensic interpretation of the results remain. For example, how long
after firing a weapon can the NG be detected?

B. Liquid Chromatography/Mass Spectrometry

The thermal instability of most explosives, along with the requirements of high sensitivity of the analysis, limits the number of analytical
methods that can be used for explosive analysis. The very nature of
explosives precludes the application of certain chromatographic procedures. High temperature gas chromatography is generally difficult to
employ since most explosives will not survive the thermal conditions
required for their vaporization and separation. Some of these difficulties
can be reduced using cold on-column capillary injection techniques and
short columns.

The majority of recently published reports dealing with the analysis of explosive material has involved high performance liquid chromatography (HPLC) with a variety of detection systems.[35] These detection systems include ultraviolet (UV) absorption,[36] electrochemistry,[37,38] thermal energy analysis,[39] and electron capture.[40]

Several LC/MS systems have been used for analysis of explosives. These systems are the LC/direct liquid introduction/MS (DLI) with either positive or negative chemical ionization (CI), and LC with off-line CI/MS detection. Sensitivity of the DLI system is limited because only 1 to 2% of the effluent enters the mass spectrometer. The off-line technique suffers from the inconvenience of sample collection and ineffective time utilization. Limits of detection in the nanogram to subnanogram range for CI/MS have been reported.[41,42]

1. Thermospray of Explosives

The thermospray (TS) LC/MS interface significantly improves the sensitivity, reliability, and applicability of LC/MS. The principles and applications of TS have been reviewed by Vestal and co-workers.[43-46] The complete or partial vaporization of a liquid as it flows through a heated capillary has been defined as thermospray. TS provides stable vaporization of a wide variety of solvent systems including volatile aqueous buffers with flow rates up to 2.0 ml/min. The TS is capable of producing intact molecular ions for many molecules which can be ionized in solution.

Recently, Voyksner and Yinon demonstrated that thermospray is applicable to the trace analysis of explosive compounds, technical explosives mixtures, and hand swabs.[47] They also observed that the gentle ionization provided by TS allows for the production of molecular or pseudomolecular ions of explosives compounds as opposed to the highly fragmented spectra observed for explosives under electron impact or chemical ionization conditions.

Berberich et al.[48] have extended the range of applications to include a number of commercial and military explosives, stabilizers in double-based smokeless powders, and postblast explosive residues. Various ionization modes were employed. The isocratic chromatographic conditions which were developed provided for a rapid separation of explosive mixtures in less than 5 min.

a. Thermospray Spectra

A Vestec thermospray source, interfaced to a Finnigan-MAT TSQ-45 triple stage quadrupole mass spectrometer, was used to analyze the explosives. The various operating conditions of the LC/TS/MS were optimized to yield the maximum sensitivity for the ions produced by LC/TS (filament-on)/MS of TNT and PETN. The greatest sensitivity for explosive compounds was obtained in the negative ion mode with filament-on ionization.[47,48]

The negative ion LC/TS (filament-on)/MS mass spectra of RDX, HMX, PETN, and tetryl are displayed in Figure 11. All spectra exhibit abundant molecular or pseudomolecular ions and little fragmentation, allowing for easy identification.

A number of compounds play a significant role in the manufacturing of smokeless and double-based smokeless powders. In contrast to the nitro-containing high explosives, this particular group of compounds produced intense positive ion spectra. The positive ion LC/TS (filament-on)/MS mass spectra of diphenylamine (DPA) and ethyl centralite were characterized by $[M + H]^+$ ions of m/z 170 and 269, respectively. The positive ion LC/TS(filament-on)/MS spectrum of monomethylaminenitrate (MMAN) had its most abundant ion at m/z 108, due to $[M + 16]^+$ and an ion at m/z 92, corresponding to $[M]^+$. A summary of the major ions of the LC/TS(filament-on)/MS mass spectra of the explosives and additives is presented in Table 2.

b.   Military and Commercial Explosives

The liquid chromatographic separation was performed using a premixed isocratic mobile phase consisting of 60/40 methanol/0.10 M ammonium acetate aqueous buffer on a C-18 column (5 mm × 15 cm, 5 μm packing).

A typical separation is shown by the reconstructed ion chromatogram (RIC) in Figure 12. The sample injected contained approximately 20 pg each of RDX, TNT, and PETN.

A number of commercial and military explosives were analyzed including military C-4 and PE-2. Tovex 300 (DuPont), a commercial blasting agent sensitized with monomethylaminenitrate (MMAN), and NONEL (Ensign Bickford), a water-resistant primer/detonating lead containing HMX and aluminum dust, were also examined. Trace amounts of HMX were found in samples of U.S.-manufactured C-4.

Hercules Red Dot, a double-based smokeless powder, was characterized by simultaneous positive and negative ion detection. The negative ion trace indicates the presence of NG, while the diphenylamine and ethyl centralite are observed only in the positive ion mode.

c.   Explosive Residue

Developing a method of analysis for identifying postblast residue is of equal utility as the characterization of unexploded materials. In order to evaluate the effectiveness of LC/TS/MS for residue analysis, a number of

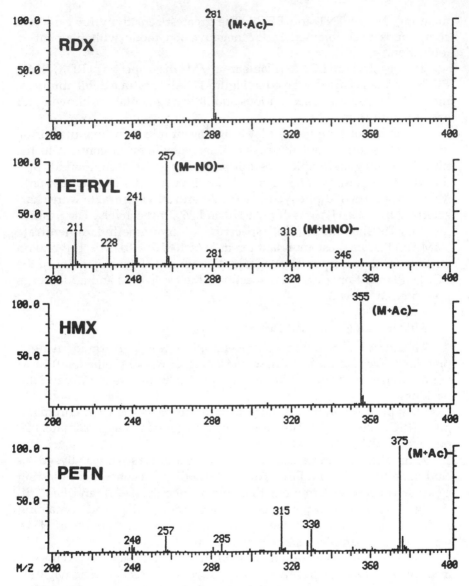

Figure 11.
Negative ion thermospray (filament-on)/mass spectra for RDX, HMX, PETN, and tetryl. (From Berberich, D. W., Yost, R. A., and Fetterolf, D. D., *J. Forensic Sci.*, 33, 946, 1988. With permission.)

improvised explosive devices were constructed and detonated under controlled conditions. For example, a propane tank recovered from near the seat of a C-4 explosion was swabbed with acetone. Extracts of the swabs showed the presence of RDX and HMX residue with selected ion monitoring.[48]

TABLE 2.

Major Ions in LC/TS (Filament-On)/MS Mass Spectra

Compound injected	Mass	Peak % RA[a]	Identity
TNT	227	100	[M]$^-$
	226	42	[M–H]$^-$
	210	51	[M–OH]$^-$
	197	15	[M–NO]$^-$
RDX	281	100	[M + CH$_3$COO]$^-$
HMX	355	100	[M + CH$_3$COO]$^-$
PETN	378[b]	100	[M + NO$_3$]$^-$
	375[b]	100	[M + CH$_3$COO]$^-$
	330	22	[M + CH$_3$COO–COOH]$^-$
	315	36	[M–H]$^-$
Tetryl	346	5	[M + CH$_3$COO]$^-$
	318	29	[M + H + NO]$^-$
	257	100	[M–NO]$^-$
	241	62	[M–NO$_2$]$^-$
NG	289	100	[M + NO$_3$]$^-$
	241	21	[M + 14]$^-$
MMAN	108	100	[M + 16]$^+$
	92	100	[M]$^+$
DPA	170	100	[M + H]$^+$
Ethyl centralite	269	100	[M + H]$^+$

[a] Relative abundance.
[b] The base peak for PETN varied between these two ions.

From Berberich, D. W., Yost, R. A., and Fetterolf, D. D., *J. Forensic Sci.*, 33, 946, 1988. With permission.

Figure 13 demonstrates the advantages offered by simultaneous positive and negative ion detection capabilities. The end cap of a pipebomb containing double-based smokeless powder was recovered from the debris and washed with acetone. The presence of DPA in the positive ion mode and the presence of NG in the negative ion traces are indicative of the use of a double-based smokeless powder.[48]

The handling of explosives or their contact with other surfaces can also leave physical trace evidence. The extract of a hand swab from an individual who handled a letter bomb explosive showed the presence of RDX and PETN.[47]

2. Electrospray Ionization

A new ionization technique known as electrospray ionization (ESI) for liquid introduction into a mass spectrometer by loop injection or by liquid

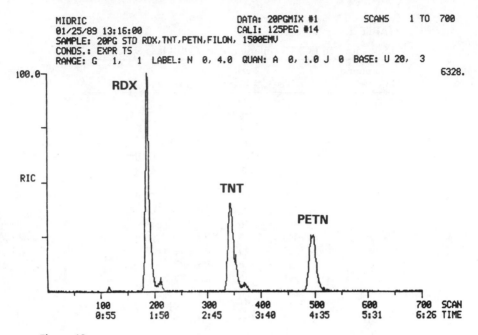

Figure 12.
LC/TS/MS reconstructed ion chromatogram for analysis of 20 pg of RDX, TNT, and PETN.

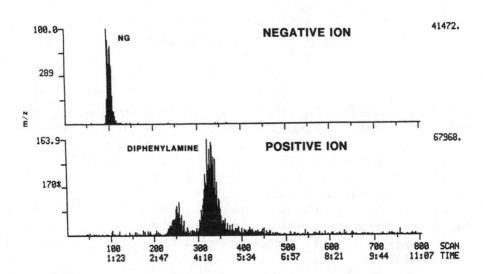

Figure 13.
Detection of NG and DPA on the extract of fragments from an improvised explosive device (pipebomb) containing double-based smokeless powder. (From Berberich, D. W., Yost, R. A., and Fetterolf, D. D., *J. Forensic Sci.*, 33, 946, 1988. With permission.)

chromatography has emerged. The fundamentals of electrospray have been reviewed by Fenn et al.[49] While ESI is generally used with large biomolecules, we have found it to be applicable to small molecules such as drugs and explosives.

In ESI, samples and mobile phase are introduced through a fused silica capillary column into the nozzle region. A potential difference of several kilovolts between the nozzle and source block results in ionization of the analyte. In the Finnigan-MAT electrospray, a nozzle/skimmer arrangement and an RF-only quadrupole focus the resulting ions into the mass analyzer. The only ion formed in the ESI of PETN, RDX, and TNT in a 0.005% HNO_3, 50% methanol/50% water mobile phase was found to be $(M + NO_3)^+$. The simplified spectra permit easy identification and are amendable to subsequent analysis by MS/MS techniques.

III. DETECTION OF HIDDEN EXPLOSIVES

A. Introduction

The basic "Ss", or figures of merit, of any analytical procedure are Speed, Sensitivity, Selectivity, and, perhaps, the most important, the cost, "$". For forensic and security screening of hidden explosives by portable instrumentation, or "sniffers", three more "Ss" must be added to the list, namely Size, Simplicity, and Sampling. A number of individuals would also add a fourth new "S", that being Skepticism. The skepticism generally arises from an incomplete understanding of the properties of explosives described in Section I and a failure to consider the other seven Ss and their relevance to explosive detection.

Terrorist events such as PAN AM 103 in December 1988, the World Trade Center bombing in New York City in February 1993, as well as the daily occurrences in South America and the Middle East, present unique forensic and analytical challenges. The detection of a few nanograms or picograms of explosive residue from debris, scattered over hundreds of square miles or in hundreds of tons of debris, is an arduous task. Clearly, sensitivity and selectivity are important. Screening debris for traces of explosive at the crime scene may provide law enforcement officers important investigative leads in a timely fashion.

The detection of explosives hidden in terrorist safe houses or conveyances, such as automobiles or aircraft, presents an equally difficult set of challenges. The collection of trace physical evidence on a suspect or on his belongings can provide evidence of probable cause for a search warrant, help identify a potential terrorist, and locate concealed explosives.

The detection of explosives hidden on passengers, carry-on and checked baggage, and cargo is compounded by the shear volume of air traffic. Detection requirements often allow only a few seconds to screen an item for hidden explosives. The systems must contain adequate selectivity to produce low false alarm rates, otherwise confidence will be lost in the detectors' ability.

The basic properties of explosives previously discussed play an important role. The most important step in the mass spectral identification of postblast residue or in the detection of hidden explosives lies not with the mass analyzer but with the sample collection. Therefore, thermal stability, absorbtivity, and vapor pressure cannot be ignored. In addition, these analyses are compounded by the exacting requirements of the legal system if successful prosecution of the criminals is to occur.

B. Sampling and Concentration

The sampling of explosives is best divided into two categories, (1) vapor sampling and (2) surface particle collection.

1. Vapor Sampling

Attempts to detect pure explosive vapor often fall short of expectations. Consider a simple scenario of a briefcase containing a small amount of plastic explosive such as C-4 (RDX). A simple calculation, using Boyle's gas laws and the equilibrium vapor pressure of pure explosives at STP, shows that a standard-sized empty briefcase (10.6 l) would contain only 6.3×10^{-11} g or 5.94×10^{-14} g/ml RDX vapor. This is, however, the best that one can expect.

In reality the amount of explosive vapor available for analysis is much less. Such a calculation fails to take into account Raoult's Law which states that the true vapor pressure of the explosive will be reduced because it is a component of a mixture. McGann and co-workers have measured two orders of magnitude difference for pure RDX and C-4.[50] Kenna and co-workers have also discussed the vapor emission of plastic explosives based upon the solubility of the explosive in the oil matrix.[51]

Another factor to be considered is the kinetics or time dependence associated with vaporization and the absorption of explosives on surfaces. TNT has been shown to vapor deposit on a variety of surfaces.[52] Griffy's kinetic model of explosive vapor concentration estimated that until a monolayer of TNT was deposited on the surrounding walls of a room, the vapor pressure was 6×10^{-4} less than the equilibrium vapor pressure.[53] Experimental verification of the model led Griffy to conclude that "The practical implication of this result is that explosive detectors which function by collecting air samples are unlikely to be effective unless their sensitivity is orders of magnitude greater than that required to detect equilibrium amounts of the vapor."[54]

Finally, the emission rate of explosives from the concealment device must be taken into account. Lucero has estimated that this is approximately 1 nl/min.[55] The emission rate of the explosive is also affected by any barrier material, (e.g., plastic wrap or foil) wrapped around it. Hobbs has estimated that only 1 to 10% of the air can be removed from a suitcase by compression.[56] Microwave irradiation of the bag followed by sampling through a flexible perforated tube into a mass spectrometer has been proposed.[57]

In the detection of explosive vapors for forensic purposes, vapor phase sampling of explosives debris has been carried out on Chromsorb 102,[58] charcoal,[59] silica gel,[60] ODS-Sherisorb and alumina,[61] and tenax[62] with elution by a variety of solvents. The adsorption characteristics of explosives on common adsorbents have been reviewed.[63]

Generally, these adsorbent traps are quite efficient in the sampling of air at flow rates of 10 to 100 ml/min with total volumes of 5 to 200 ml.[64] While this may be satisfactory for postblast vapor analysis in the laboratory, these adsorbents are incompatible with the need to sample a passenger-screening portal of 80 ft^3 in a few seconds. A recent review also points out the general requirements and concerns associated with the conditioning, reconditioning, analyte breakthrough, and possible interferant introduction.[65] In addition, the wide variety of hydrocarbons, aromatics, and substituted aromatics associated with the baggage and security areas, ramps, and ticket counters of airports[66] may quickly saturate the adsorbents.

2. Surface Particle Collection

Contamination of the hands has been shown to occur after handling commercial[67] and military explosives.[68] The persistence of these explosives on hands and evidence of contact transfer to other surfaces is well documented in forensic sciences.[69,70] In a surface-testing protocol developed by Neudorfl and co-workers,[71] a series of thumbprints was made on ten consecutive stainless steel plates after touching SEMTEX. The amount of RDX transferred ranged from 4495 ng on the first thumbprint to 135 ng of RDX on the tenth thumbprint.[71] Clearly, these amounts are many orders of magnitude greater than that which would be expected from the pure explosive vapor discussed above. Such amounts can be readily detected using a simple colorimetric chemical test kit.[72] For most mass spectrometers, 135 ng of an analyte would be detectable if the sample could be collected and transferred into the ion source. Clearly the sampling "S" is the major obstacle to the analysis of hidden explosives by mass spectrometry.

A method of detecting contraband (explosives, drugs, firearms) in cargo containers has been proposed in which the container is transported over "speed bumps" to disturb particulate matter.[73] The particulates may be the explosives themselves or explosive vapors adsorbed on atmospheric dust particulates. The samples can then be collected on a filter medium coated with an organosilicon chromatographic phase polymer oil.[73]

C. Ion Mobility Spectrometry

1. Introduction

Ion mobility spectrometry (IMS) was first introduced by Cohen and Karasek in 1970.[74] Much of the earlier work on the applications of IMS technology focused on fundamental studies and laboratory feasibility. In a recent review, it was pointed out that IMS technology is experiencing a resurgence of interest in specific purpose detection systems because of its analytical flexibility.[75] Since its inception, the potential of IMS for explosives detection was noted with detection limits for TNT of 0.01 ppb being reported.[76] Forensic applications of IMS technology from 1970 to 1989 have been thoroughly reviewed by Karpas.[77]

The ease of use and portability permit operation in real world scenarios not only as a forensic tool but also as an investigative tool. The value of IMS as an investigative tool (rather than purely a laboratory tool) is exemplified in recent applications which have included the detection of drug microparticulates on hands,[78,79] determining cocaine in injection-molded plastic,[80,81] and use in customs scenarios.[82]

The IMS consists of two main areas: the reaction region and the drift region. In the reaction region, atmospheric pressure carrier gas (purified air), the reactant gas, and any internal calibrant are ionized by a ^{63}Ni beta emitter to form negative ions. The ionization of explosives under a wide variety of conditions[77] and the use of modified ion chemistry to improve the sensitivity and selectivity of detection by IMS have been discussed by Danylewich-May.[83] In general, negative ion detection provides a high degree of selectivity because only highly electronegative compounds such as explosives are ionized.

Under the influence of an electric field, the mixture of reactant and product ions reaches a shutter grid that separates the reaction region and the drift region. The shutter grid is made of sets of thin mesh wires with a voltage bias between them. When the shutter grid is "on" (with bias voltage applied), the ions are attracted to the gating grid and lose their charge. For a brief amount of time the grid is turned "off". Ions are then transmitted into the drift region of the cell. In the drift region an electric field gradient is applied. The ions migrate through the electric field, but at the same time are hindered by the countercurrent drift gas. The smaller, compact ions have a higher mobility than the heavier ions, and therefore traverse the region and collide with the electrometer plate in a shorter time. With the aid of a microprocessor, a plot of ion current intensity vs. the time elapsed from the opening of the shutter grid gives the mobility spectrum or plasmagram.

In general, for a given temperature, T (Kelvin), of the drift gas and pressure, P (torr), the mobility of an ion is given as reduced mobility, K_0, in the form of

$$K_0 = [d/(Et)] \quad (273/T) \quad (P/760) \tag{1}$$

where d is the length of the drift region, E is the electric field strength in V/cm, and t the mobility time in seconds.

2. IMS Spectra and Limits of Detection

The forensic applications of ion mobility spectrometry for the detection of physical trace evidence from a pre- and postblast explosives incidents have recently been reported.[84] Using a Barringer Instrument IONSCAN Model 250, the drift times, reduced mobility (referenced to TNT, K_0 = 1.451), and the limit of detection for these five common explosive components were recorded and are shown in Table 3. The detection of a specific ionic species on the computer-displayed plasmagram was found to range from 200 pg to 80 ng. The formation of multiple species greatly increases the specificity.

TABLE 3.

IMS Ion Characteristic for Some Common Explosives

Peak	Proposed species (−)	Mass	Drift time (ms)	K_0 (cm²V⁻¹S⁻¹)	L.O.D.
TNT	TNT–H	227	14.52	1.451	200 pg
RDX-1	RDX + Cl	257	15.19	1.387	200 pg
RDX-2	RDX + NO₃	284	16.03	1.314	800 pg
RDX-3	RDX + (RDX + Cl)	479	22.22	0.948	1 ng
PETN-1	PETN–H	316	17.37	1.213	80 ng
PETN-2	PETN + Cl	351	18.40	1.145	200 pg
PETN-3	PETN + NO₃	378	19.08	1.104	1 ng
NG-1	NG + Cl	262	15.73	1.339	50 pg
NG-2	HG + NO₃	289	16.50	1.275	200 pg
NO₃	H₂O + NO₃	100	10.93	1.927	200 pg

From Fetterolf, D. D. and Clark, T. D., J. Forensic Sci., 38, 28, 1993. With permission.

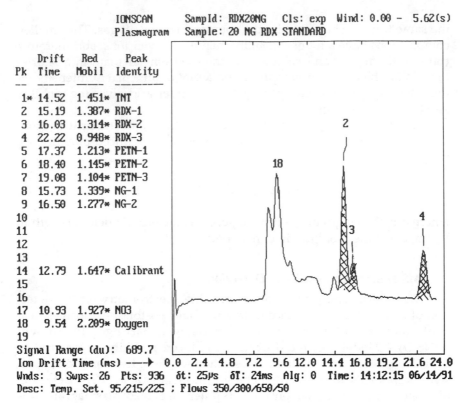

```
              IONSCAN      SampId: RDX20NG   Cls: exp   Wind: 0.00 -  5.62(s)
              Plasmagram   Sample: 20 NG RDX STANDARD

      Drift  Red     Peak
  Pk  Time   Mobil   Identity
  --  ----   ----    --------
  1*  14.52  1.451*  TNT
  2   15.19  1.387*  RDX-1
  3   16.03  1.314*  RDX-2
  4   22.22  0.948*  RDX-3
  5   17.37  1.213*  PETN-1
  6   18.40  1.145*  PETN-2
  7   19.08  1.104*  PETN-3
  8   15.73  1.339*  NG-1
  9   16.50  1.277*  NG-2
  10
  11
  12
  13
  14  12.79  1.647*  Calibrant
  15
  16
  17  10.93  1.927*  NO3
  18   9.54  2.209*  Oxygen
  19
  Signal Range (du): 689.7
  Ion Drift Time (ms) ---->  0.0  2.4  4.8  7.2  9.6 12.0 14.4 16.8 19.2 21.6 24.0
  Wnds:  9 Swps: 26  Pts: 936  δt: 25μs  δT: 24ms  Alg: 0  Time: 14:12:15 06/14/91
  Desc: Temp. Set. 95/215/225 ; Flows 350/300/650/50
```

Figure 14.
Plasmagram of 20 ng of RDX. (From Fetterolf, D. D. and Clark, T. D., *J. Forensic Sci.*, 38, 28, 1993. With permission.)

It can be seen in Table 3 that more than one species is generated for some of the explosives, depending upon the amount of explosive present. For example, three species labeled RDX-1, RDX-2, and RDX-3 are formed in RDX. It is theorized that these species are the chloride adduct $(RDX + Cl)^-$, the nitrate adduct $(RDX + NO_3)^-$, and an adduct between a neutral RDX molecule and the chloride adduct $(RDX + (RDX + Cl))^-$. At 200 pg of RDX in the ionization region, only the chloride adduct is formed. As the concentration increases to 1 ng, more free neutral RDX molecules are present which can undergo ion molecule reactions with the chloride adduct. Proposed ionic species for the other explosives are presented as well. The plasmagram of RDX is shown in Figure 14.

3. Applications

a. Surface Contamination

To demonstrate the transfer of explosives from hands to surfaces a subject touched C-4 (RDX). The subject then enacted several normal stages

TABLE 4.

Contact Transfer of C-4 from Hands to
Automobile Surfaces

Car area	Before touching	After C-4 transfer
Hands	−	+
Hood	−	+
Door handle	−	+
Hatchback	−	+
Steering wheel	−	+
Gear shift	−	+
Keys	−	+

From Fetterolf, D. D. and Clark, T. D., *J. Forensic Sci.*,
38, 28, 1993. With permission.

in operating a car, including: opening and closing the hood, the driver side door, the trunk, and handling the steering wheel, gearshift, and keys.

Samples were collected before and after the contact transfer from each of the touched areas of the car and from the subject's hands by vacuuming onto the teflon disk. As shown in Table 4, the car surfaces and hands were negative for RDX prior to touching the explosive. After contact, all touched areas showed easily detectable RDX residue. In a separate experiment, after handling C-4, eight consecutive hand washings with soap and water were required before the IMS could no longer detect the RDX. Swabbing or wiping the surface with the filter disk is also a suitable means of collection. Other surface sampling/collection devices include a heated sampling cone[85] and the use of laser desorption to volatilize the explosives on the surface[86] prior to ion mobility spectrometry.

b. Postblast Residue

Following a terrorist attack, rapid and accurate analysis of postblast residues plays a vital role in the bombing investigation. The first question often posed by an investigator following a blast is "What was the explosive?". The analysis may provide the link between a suspect and the type of explosive used. Yelverton has demonstrated the detection of postblast RDX vapor using a quartz tube preconcentrator and IMS.[87] Because of its portability, the IMS can be taken to a bombing crime scene for preliminary analysis to aid law enforcement investigators.

A number of improvised explosive devices (pipe bombs) were prepared. The pipes contained Hercules Green Dot, Royal Scott, Hercules Red Dot, and Winchester Ball double-based smokeless powder, Pyrodex (a black powder substitute), and black powder. Following detonation, fragments of the pipes were recovered for analysis. A single fragment, just a few square inches in size, was vacuumed. Fragments from the four pipes

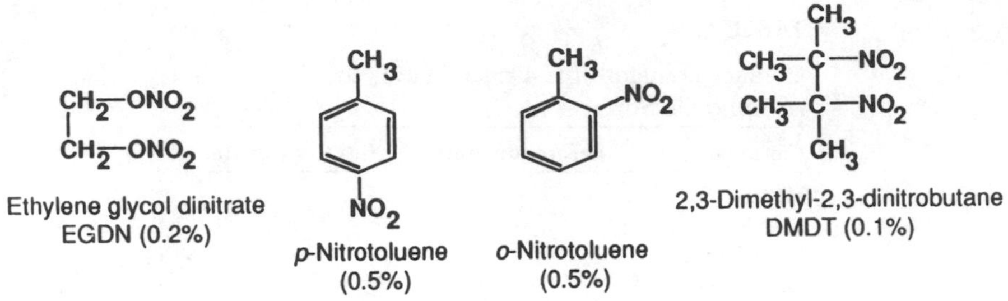

Figure 15.
Structure of proposed ICAO explosive taggants.

containing the double-based smokeless powder alarmed positive for nitroglycerin.[84]

Postblast explosives' vapors have also been observed to be collected on nearby surfaces. For example, NG residue has been detected on wipe samples taken from a night deposit box following an explosion. Vacuum samples from the inside, outside, and handles of a gym bag believed to be used to transport the explosives to the bank were also positive for NG.

Fabric appears to be a good preconcentrator of explosives' vapors. A demolition block (1.25 lb) of C-4 was placed in a suitcase containing ten clothing items. Before the blast, the clothing and suitcase were clear of explosives. The C-4 was detonated and the postblast debris was collected. Postblast RDX residue was detected on all articles of clothing using the vacuum sample method.[84]

c. Explosive Taggants

The International Civil Aviation Organization (ICAO) has adopted an international convention that will enhance the detectability of explosives by adding volatile markers to them.[88] The explosives covered by the convention include RDX, PETN, and HMX. According to the convention, current stocks of unused explosives must be used or destroyed by military and police. The markers to be considered are shown in Figure 15 and include ethylene glycol dinitrate (EGDN), o-nitrotoluene, p-nitrotoluene, and 2,3-dimethyl-2,3-dinitrobutane (DMDT) at 0.1 to 0.5%. The markers were detected at the 0.1 to 1 ng in an IMS optimized for their detection.[89]

d. Vapor Generator Calibration

A novel application of the IMS was the use of a PCP Model 110 to calibrate a vapor generator.[90] The vapor generator is designed to be used to verify and calibrate explosive vapor detection systems. The explosive vapors generated from an isothermal reservoir were collected on a quartz wool preconcentrator tube and compared with an external calibration curve for the IMS. The IMS was calibrated by using standard solutions of the explosives. The vapor generator produced a 5-s wide pulse of explosive vapors ranging from 50 to 1350 pg depending on temperature and explosive.

e. Investigative Support

Because of their size, portability, and ease of use, IMS detectors have been employed in direct support of field investigations involving the search for hidden explosives.[91]

A briefcase-sized Graseby Dynamics PD-5 was found to be extremely sensitive for the higher vapor pressure-nitrated ester explosives (NG and EGDN). The briefcase IMS was found to provide a better combination of sensitivity, false alarm rate, and speed than electron capture-based detectors designed for the same purpose.[92] In our trials, a quarter stick of dynamite (0.25 lb) could easily be detected in the trunk of a car or hidden in a package in an auditorium with less than 1 h soak time.

Following a domestic dispute, a local police agency requested the assistance of the FBI Laboratory in locating and identifying homemade nitroglycerin in a basement.[91] The PD-5 was used to locate the two bottles of hidden NG. The NG was rendered safe by the bomb squad. In a similar incident the PD-5 was used to search 1300 rental storage lockers and six apartment buildings for a cache of NG explosives believed to be in the possession of a known drug dealer.[91]

The heightened awareness of potential terrorist threats during the Gulf War required extraordinary security measures at Superbowl XXV in Tampa, FL. Metal detectors were used to screen fans at the stadium gates, and all portable electronic devices were prohibited. A number of IMS and chemiluminescence systems were employed to search vehicles and stadium locations inaccessible to the bomb-sniffing canines. RDX residue was detected on the trunk of a car used to store the canine training aids.[91] Traces of explosives were also found in locations where the canine training aids had been hidden and removed.

D. Tandem Mass Spectrometry

The potential role of a portable computer-controlled tandem mass spectrometry in hidden explosives detection was first noted by Yinon.[93] Two different ionization modes have been found to be sensitive and selective for detecting hidden explosives, namely atmospheric sampling glow discharge ionization (ASGDI) and atmospheric pressure ionization (API). The ability of both of these ion sources to directly sample ambient air into a mass spectrometer makes them attractive for the rapid and direct analysis of explosives. The negative-ion modes employed reduce interference from potential background material.

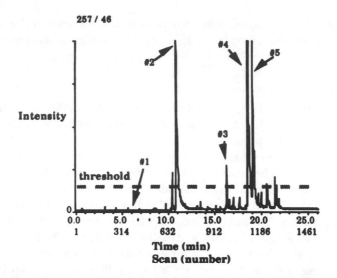

Figure 16.
Detection of RDX on suitcases by selected reaction monitoring. (From Davidson, W. R., Stott, W. R., Akery, A. K., and Sleeman, R., Proc. First Int. Symp. on Explosive Detection Technology, Atlantic City, NJ, 1991, 663. With permission.)

1. Atmospheric Pressure Ionization (API)

API has been reviewed in depth[94] and has provided picogram[95] and subpicogram[96] detection of a wide variety of compounds. Using API and a Sciex TAGA 2000 quadrupole mass spectrometer, TNT (0.5 ng/m^3) and dynamite were detected in the cabin of a DC-8 aircraft.[97] A commercial API/MS/MS instrument known as the CONDOR is marketed by British Aerospace/SCIEX. It has been shown that modification of the ionization process by the formation of Cl adducts leads to an enhancement in sensitivity of a factor of 10.[98] Selected reaction monitoring of the daughter ions from (EGDN + Cl)$^-$ resulted in the detection of 600 ppt of EGDN using a real-time continuous action preconcentrator.[99] API/MS/MS has been reported for the real time detection of NG, DNT, and TNT.[100]

Figure 16 shows the detection of RDX residue on four suitcases in an aviation security demonstration.[101] Five suitcases were prepared with traces of explosives. Peaks labeled 2 through 5 show the detection of RDX on the suitcases, above the operators' alarm threshold, as numerous bags passed by the real-time sampling system. Because of sample throughput limitations with a prototype, real-time sampling system bag number 1 was not sampled. The use of multiple reaction monitoring, in which several parent and daughter ion combinations were monitored for the detection of RDX from SEMTEX, on a suitcase in aviation security has also been reported.[102]

2. Atmospheric Sampling Glow Discharge Ionization (ASGDI)

Atmospheric sampling glow discharge ionization, developed at Oak Ridge National Laboratory, has been shown to be a sensitive means of direct sampling and ionizing trace organic species in ambient air.[103,104] The source consists of a pair of parallel discharge plates in the ionization region and an ion exit aperture. The source is evacuated to about 0.8 torr with a sample flow rate of 5 ml/s using a 15-l/s mechanical pump. By applying a potential difference of about 400 V between the plates or apertures, a glow discharge is produced. The versatility of the source can be seen in that it has been interfaced to three different MS/MS instruments: a quadrupole/time of flight (QT), an ion trap mass spectrometer (ITMS), and a triple stage quadrupole (TSQ). Another attractive feature is the ability of the source to operate unattended for months at a time.

a. Quadrupole/Time of Flight

The ASGDI source interfaced to a quadrupole time of flight (QT) tandem mass spectrometer has been described.[105] ASGDI negative-ion detection of TNT at the 1 to 2 ppt level with six orders of linear dynamic range has been reported.[103] Samples are collected remotely using a quartz wool preconcentrator tube. An independent evaluation of the sensitivity and selectivity of the detector was carried out by Conrad.[106] The detector unambiguously identified bomb quantities of TNT, C-4, tetryl, PETN, and HMX and a number of propellants. The detection of C-4 carried on an individual and hidden in mail was demonstrated. RDX was detected at levels as low as 0.3 ppt when operated in the targeted daughter ion (selected reaction monitoring mode). The detector was shown to be free of interference from solvent vapors and commonly encountered items such as food, tobacco products, fuel, and perfume.

b. Ion Trap Mass Spectrometry (ITMS)

A promising MS/MS explosives detector is based upon the ion trap detector.[107] In comparison to the QT, the ion trap is smaller and offers improved daughter ion resolution and conversion efficiency. The negative ion water chemical ionization (OH^- and O^- reagent ions) MS/MS daughter ion spectra of TNT and RDX has been reported.[108] The negative ion formation of TNT and RDX by the use of reagent anion injection from an external source into an ITMS has been described.[109] The detection of less than 6 ppb of TNT in 0.35 s[110] has been reported. Longer sampling and ion storage times could dramatically improve this limit of detection. The collisional activation of the molecular anion of TNT in an ion trap has been carried out using random noise applied to the endcaps.[111]

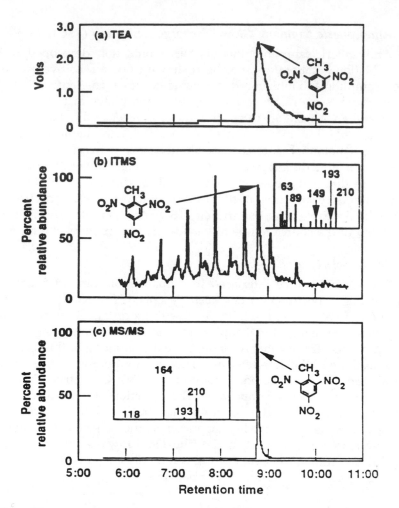

Figure 17.
GC/TEA/ITMS chromatographic profiles of TNT spiked diesel fuel: (a) TEA profile, (b) ITMS total ion current profile, and (c) selected reaction monitoring. (From Alcaraz, A., Martin, W., and Andresen, B. D., Proc. 39th ASMS Conference on Mass Spectrometry and Allied Topics, Nashville, TN, 1991, 158. With permission.)

The ITMS has also been coupled to a gas chromatograph-thermal energy analyzer (TEA) by means of a capillary column splitter.[112] The TEA is a nitro-specific chemiluminescence detector with picogram limits of detection for explosives which is widely used in forensic laboratories.[113] A transportable TEA system known as the Thermedics EGIS has also been developed for the detection of trace explosive residue.[114] Figure 17 shows the chromatographic profile obtained with the TEA analyzer, the ITMS total ion current profile, and the resulting selected reaction monitoring trace (SRM) for the reaction m/z 210 → m/z 164 for a sample of diesel fuel spiked with TNT.[112]

This combination may well represent the most powerful combination of selectivity, sensitivity, and speed available with today's technology. False alarms should be reduced to a bare minimum with the combination of retention time, chemiluminescence behavior, and selected reaction monitoring of the ITMS.

c. Triple Stage Quadrupole (TSQ)

The ASGDI source has also been interfaced to a commercial Finnigan-MAT TSQ-700 triple stage quadrupole mass spectrometer.[115] A separate flange assembly and minor lens stack modifications were required. Such a combination provides the forensic mass spectroscopist with another tool for the identification of explosives.

The ASGDI negative ion spectra of TNT, PETN, and RDX are shown in Figure 18.[116] Only TNT exhibits a molecular anion at m/z 227. PETN is characterized by major fragments at m/z 46 and m/z 62. RDX exhibits loss of NO_2 and fragments associated with ring opening. A comparison of the daughter ion spectra of m/z 227 from TNT is shown in Figure 19.[116] The higher fragmentation efficiency of the ITMS over the TSQ is exhibited by the nearly complete conversion of M⁻ (m/z 227) to m/z 210 (M–OH)⁻.

IV. CONCLUSIONS

The future of mass spectrometry in the analysis and detection of explosives appears bright, provided one pays strict attention to the combined fundamental properties of explosives: absorbtivity, electronegativity, frangibility, thermal stability, and low vapor pressure.

Dramatic developments in recent years in smaller, computerized, and ruggedized IMS systems have paved the way for continued forensic application in the field. Reducing the amount of debris which must be forwarded to the laboratory for more thorough analysis is crucial. Providing the investigator with time-sensitive information can make his job easier and lead to quicker apprehension of the suspect.

Fixed-site aviation security applications demand real time, high sensitivity, high throughput methods of sample collection, and introduction. Tandem mass spectrometry coupled with other detection systems such as chemiluminescence may provide unparalleled degrees of specificity leading to acceptable false alarm rates.

The requirements of the judicial process demand the precise identification of physical trace evidence from explosives, if successful prosecution of the suspect parties is to be carried out. Such demands include calibra-

MS of TNT

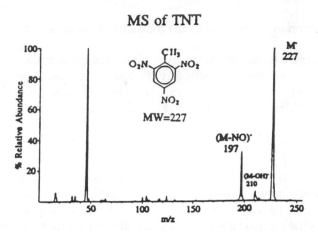

MS of PETN

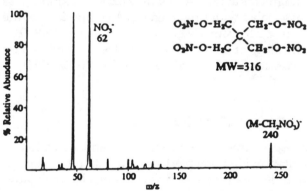

MS of RDX

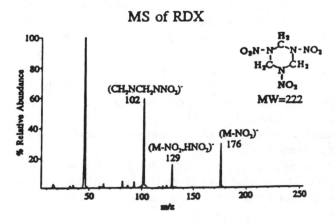

Figure 18.
ASGDI mass spectra of TNT, PETN, and RDX. (From Glish, G. L., McLuckey, S. A., Grant, B. C., and McKown, H. S., Proc. First Int. Symp. on Explosive Detection Technology, Atlantic City, NJ, 1991, 642. With permission.)

MS/MS of TNT (m/z 227) vs Instrument

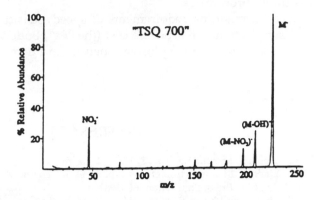

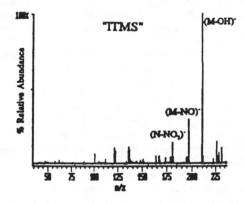

Figure 19.
Comparison of daughter spectra from an ASGDI source interfaced to various MS/MS instruments: (a) QT, (b) ITMS, and (c) TSQ 700. (From Glish, G. L., McLuckey, S. A., Grant, B. C., and McKown, H. S., Proc. First Int. Symp. on Explosive Detection Technology, Atlantic City, NJ, 1991, 642. With permission.)

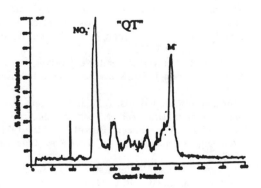

tion, good laboratory practice, and documented quality control/quality assurance procedures.

The demanding requirements of speed, sensitivity, selectivity, size, and simplicity at a reasonable cost (the "$s") bode well for the future of mass spectrometry in explosives analysis and detection.

REFERENCES

1. Yinon, J., *Forensic Sci. Rev.*, 3, 17, 1991.
2. Yinon, J., Mass spectrometry of explosives, in *Forensic Mass Spectrometry*, Yinon, J., Ed., CRC Press, Boca Raton, FL, 1987.
3. Yinon, J. and Zitrin, S., *Modern Methods and Applications in Analysis of Explosives*, John Wiley & Sons, Chichester, 1993.
4. Beveridge, A. D., *Forensic Sci. Rev.*, 4, 18, 1992.
5. Lloyd, J. B. F., HPLC of explosives materials, in *Advances in Chromatography*, Giddings, J. C., Grushka, E., and Brown, P. R., Eds., Marcel Dekker, New York, 1991.
6. Urbanski, T., *Chemistry and Technology of Explosives*, Vol. 1–4, Pergamon Press, New York, 1988.
7. Meyer, R., *Explosives*, VCH Publishers, New York, 1987.
8. Conrad, F. J., Proceedings of 25th Annual Nuclear Materials Management Mtg., Columbus, OH, 1984.
9. Peterson, P. K., Proceedings of the Int. Symp. on the Analysis and Detection of Explosives, Quantico, VA, 1983, 391.
10. Farber, M., *Mass Spectrom. Rev.*, 11, 137, 1992.
11. Jyothi Basu, V. C., Chaudhri, M. M., and Housden, J., *J. Materials Sci.*, 26, 2199, 1991.
12. Dionne, B. C., Rounbehler, D. P., Achter, E. K., Hobbs, J. R., and Fine, D. H., *J. Energ. Materials*, 4, 447, 1986.
13. Yinon, J., *Mass Spectrom. Rev.*, 1, 257, 1982.
14. Chutjian, A., Boumsellek, S., and Alajajian, S. H., Proc. First International Symposium on Explosive Detection Technology, Atlantic City, NJ, 1991, 571.
15. Reutter, D. J., Bender, E. C., and Rudolph, T. L., Proc. Int. Symp. on the Analysis and Detection of Explosives, Quantico, VA, 1983, 149.
16. Zitrin, S., *J. Energ. Materials*, 4, 199, 1986.
17. Tamiri, T. and Zitrin, S., *J. Energ. Materials*, 4, 215, 1986.
18. Tamiri, T., Zitrin, S., Abramovich-Bar, S., Bamberger, Y., and Sterling, J., GC/MS analysis of PETN and NG in post-explosion analysis, in *Advances in Analysis and Detection of Explosives*, Yinon, J., Ed., Kluwer Academic Publishers, Dordrecht, The Netherlands, 1993, 323.
19. Hobbs, J. R., Analysis of Semtex explosives, in *Advances in Analysis and Detection of Explosives*, Yinon, J., Ed., Kluwer Academic Publishers, Dordrecht, The Netherlands, 1993, 409.
20. Murrmann, R. P., Jenkins, T. F., and Leggett, D. C., *Composition and mass spectra of impurities in military grade TNT vapors*, U.S. Army Cold Regions Research and Engineering Laboratory, Special Report 158, May 1971.
21. Weinberg, D. S. and Hsu, J. P., *J. High Resol. Chromatogr. Chromatogr. Commun.*, 6, 404, 1983.
22. McCord, B. R. and Whitehurst, F. W., *J. Forensic Sci.*, 37, 1574, 1992.
23. Yost, R. A. and Fetterolf, D. D., *Mass Spectrom. Rev.*, 2, 1, 1983.

24. Busch, K. L., Glish, G. L., and McLuckey, S. A., *Mass Spectrometry/Mass Spectrometry*, VCH Publishers, New York, 1988.
25. Proceedings of the Eighth International Conference on Terrorist Devices and Methods, Washington, D.C., June 22–26, 1987 (official use only).
26. Emerson, S. A. and Del Sesto, C., *Terrorist: The Inside Story of the Highest-Ranking Iraqi Terrorist Ever to Defect to the West*, Villiard Books, New York, 1991.
27. *1991 Bomb Summary*, Bomb Data Center, FBI Laboratory, 1991, 1.
28. Wallace, C. L. and Midkiff, C. R., Smokeless powder characterization: an investigative tool in pipe bombings, in *Advances in Analysis and Detection of Explosives*, Yinon, J., Ed., Kluwer Academic Publishers, Dordrecht, The Netherlands, 1993, 29.
29. Martz, R. M., Munson, T. O., and Lasswell, L. D., Proc. Int. Symp. on the Analysis and Detection of Explosives, Quantico, VA, 1983, 149.
30. Lundy, D. R. and Kinard, W. D., ACS Symposium Series 13, American Chemical Society, Washington, D.C., 1975, 97.
31. Wolten, G. M., Nesbitt, R. S., Calloway, A. R., Loper, G. L., and Jones, P. F., *Final Report on Particle Analysis for Gunshot Residue Detection*, The Aerospace Corporation, Washington, D.C., September 1977.
32. Mach, M. H., Palloa, A., and Jones, P. F., *J. Forensic Sci.*, 23, 433, 1978.
33. Lloyd, J. B. F., *J. Energ. Materials*, 4, 239, 1984.
34. Fetterolf, D. D. and Koon, R. K., Proc. 36th ASMS Conference on Mass Spectrometry and Allied Topics, San Francisco, CA, 1988, 805.
35. Lloyd, J. B. F., HPLC of explosives materials, in *Advances in Chromatography*, Giddings, J. C., Grushka, E., and Brown, P. R., Marcel Dekker, New York, 1991.
36. Meier, E. P., Taft, L. G., Graffeo, A. P., and Stanford, T. B., Proc. Fourth Joint Conference on Sensing of Environmental Pollutants, New Orleans, LA, 1977, 487.
37. Bratin, K., Kissinger, P. T., Briner, R. C., and Bruntlett, C. S., *Anal. Chim. Acta*, 130, 295, 1981.
38. Krull, I. S. and Camp, M. J., *Am. Lab.*, 10, 63, 1980.
39. Fine, D. H., Yu, W. C., Goff, E. U., Bender, E. C., and Reutter, D. J., *J. Forensic Sci.*, 29, 732, 1984.
40. Chamberlain, A. T. and Marlow, J. S., *J. Chromatogr. Sci.*, 15, 29, 1977.
41. Vouros, P., Peterson, B. A., Colwell, L., and Karger, B. L., *Anal. Chem.*, 49, 1039, 1977.
42. Yinon, J. and Hwang, D. G., *J. Chromatogr.*, 268, 45, 1983.
43. Vestal, M. L., *Int. J. Mass Spectrom. Ion Phys.*, 46, 193, 1983.
44. Blakley, C. R., Vestal, M. L., and Carmody, J. J., *Anal. Chem.*, 52, 1636, 1980.
45. Blakley, C. R. and Vestal, M. L., *Anal. Chem.*, 55, 750, 1983.
46. Garteiz, D. A. and Vestal, M. L., *Liq. Chromatogr.*, 3, 334, 1983.
47. Voyksner, R. D. and Yinon, J., *J. Chromatogr.*, 354, 393, 1986.
48. Berberich, D. W., Yost, R. A., and Fetterolf, D. D., *J. Forensic Sci.*, 33, 946, 1988.
49. Fenn, J. B., Mann, M., Meng, C. K., Wong, S. F., and Whitehouse, C. M., *Mass Spectrom. Rev.*, 9, 37, 1990.
50. McGann, W., Jenkind, A., and Ribeiro, K., Proc. First Int. Symp. on Explosive Detection Technology, Atlantic City, NJ, 1991, 518.
51. Kenna, B. T., Conrad, F. J., and Hannum, D. W., Proc. First Int. Symp. on Explosive Detection Technology, Atlantic City, NJ, 1991, 510.
52. Bender, E., Hogan, A., Leggett, D., Miskolczy, G., and MacDonald, S., *J. Forensic Sci.*, 37, 1673, 1992.
53. Griffy, T. A., Proc. 3rd Int. Symp. on Analysis and Detection of Explosives, Mannheim-Neuostheim, Germany, 1989, 38–1.
54. Griffy, T. A., A model of explosive vapor concentration II, in *Advances in Analysis and Detection of Explosives*, Yinon, J., Ed., Kluwer Academic Publishers, Dordrecht, The Netherlands, 1993, 503.
55. Lucero, D. P., *J. Test. Eval.*, 13, 222, 1985.

56. Hobbs, J. R., Proc. New Concepts Symp. and Workshop on Detection and Identification of Explosives, Reston, VA, 1978, NTIS No. PB 296055.
57. Bather, J. M., Method and Apparatus for Detecting Dangerous Substances, International Patent Application PCT/GB86/00505, 1986.
58. Chrostowski, J. E., Holmes, R. N., and Rehn, B. W., J. Forensic Sci. Soc., 21, 611, 1976.
59. Prime, R. J. and Krebs, J., Can. Soc. Forensic Sci., 17, 35, 1984.
60. Yip, I. H. L., Can. Soc. Forensic Sci., 15, 87, 1982.
61. Lloyd, J. B. F., J. Chromatogr., 261, 391, 1983.
62. Lloyd, J. B. F., J. Chromatogr., 328, 145, 1975.
63. Wardleworth, D. F. and Ancient, S. A., Proc. Int. Symp. on the Analysis and Detection of Explosives, Quantico, VA, 1983, 391.
64. Poole, C. F. and Schuette, C. F., J. High Resol. Chromatogr., 6, 526, 1983.
65. Nunez, A. J., Gonzalez, L. F., and Janak, J., J. Chromatogr., 300, 127, 1984.
66. Jarke, F. H., Dravnieks, A., Grove, E. L. et al., Proc. New Concepts Symp. and Workshop on Detection and Identification of Explosives, Reston, VA, 1978, NTIS No. PB 296055.
67. Twibell, J. D., Home, J. M., Smalldon, K. W., and Higgs, D. G., J. Forensic Sci., 27, 783, 1982.
68. Lloyd, J. B. F. and King, R. M., J. Forensic Sci., 29, 284, 1984.
69. Lloyd, J. B. F. and King, R. M., Proc. 3rd Int. Symp. on Analysis and Detection of Explosives, Mannheim-Neuostheim, Germany, 1989, p. 9–1.
70. Lloyd, J. B. F., J. Forensic Sci. Soc., 26, 341, 1986.
71. Neudorfl, P. and McCooeye, Elias, L., Testing protocol for surface-sampling detectors, in Advances in Analysis and Detection of Explosives, Yinon, J., Ed., Kluwer Academic Publishers, Dordrecht, The Netherlands, 1993, 373.
72. Almog, J., Kraus, S., and Glattstein, B., J. Energ. Materials, 4, 159, 1986.
73. Reid, N. M. and Davidson, W. R., Method and Apparatus for Detecting a Contraband Substance, U.S. Patent 4,718,268, 1988.
74. Cohen, M. J. and Karasek, F. W., J. Chromatogr. Sci., 8, 330, 1970.
75. Hill, H. H., Siems, W. F., and St. Louis, R. H., Anal. Chem., 62, 1201, 1990.
76. Karasek, P. A., Res. Dev., 25, 32, 1974.
77. Karpas, Z., Forensic Sci. Rev., 1, 104, 1989.
78. Lawrence, A. H., Forensic Sci. Int., 34, 73, 1987.
79. Nanji, A. A., Lawrence, A. H., and Mikhael, N. Z., Clin. Toxicol., 25, 501, 1987.
80. Fetterolf, D. D., Donnelly, B., and Lasswell, L., Proc. Int. Symp. on the Forensic Aspects of Trace Evidence, Quantico, VA, in press.
81. Cocaine smuggled as ingredient in plastic, Chem. Eng. News, July 8, 1991.
82. Chauhan, M., Harnois, J., Kovar, J., and Pilon, P., Can. Soc. Forensic Sci. J., 24, 43, 1991.
83. Danylewich-May, L. L., Proc. of the First Int. Symp. on Explosive Detection Technology, Atlantic City, NJ, 1991, 672.
84. Fetterolf, D. D. and Clark, T. D., J. Forensic Sci., 38, 28, 1993.
85. Spangler, G. E., Carrico, J. P., and Kim, S. H., Proc. Int. Symp. on the Analysis and Detection of Explosives, Quantico, VA, 1983, 267.
86. Huang, S. D., Kolaitis, L., and Lubman, D. M., Appl. Spectrosc., 41, 1371, 1987.
87. Yelverton, B. J., J. Energ. Materials, 6, 73, 1988.
88. No author, Chem. Eng. News, March 4, 1991, 4.
89. Danylewich-May, L. L. and Cumming, C., Explosive and taggant detection with Ionscan, Advances in Analysis and Detection of Explosives, Yinon, J., Ed., Kluwer Academic Publishers, Dordrecht, The Netherlands, 1993, 385.
90. Davies, J. P., Blackwood, L. G., Davis, S. G., Goodrich, L. D., and Larson, R. A., Design and calibration of pulsed vapor generators for TNT, RDX and PETN, Advances in Analysis and Detection of Explosives, Yinon, J., Ed., Kluwer Academic Publishers, Dordrecht, The Netherlands, 1993, 513.

91. Fetterolf, D. D. and Whiterhurst, F. W., Proc. 39th ASMS Conference on Mass Spectrometry and Allied Topics, Nashville, TN, 1991, 1207.
92. Sheldon, T., Proc. 3rd Int. Symp. on Analysis and Detection of Explosives, Mannheim-Neuostheim, Germany, 1989, 20–1.
93. Yinon, J., Proc. Int. Symp. on the Analysis and Detection of Explosives, Quantico, VA, 1983, 1.
94. Carroll, D. I., Dzidic, I., Stilwell, R., and Horning, E., NBS Special Publication 519, Trace Organic Analysis, A New Frontier in Analytical Chemistry, 1979, 655.
95. Horning, E. C., Horning, M. G., Carrol, D. I., Dzidic, C. I., and Stillwell, R. N., *Anal. Chem.*, 45, 936, 1973.
96. Carrol, D. I., Dzidic, I., Horning, M. G., and Horning, E. C., *Anal. Chem.*, 46, 706, 1974.
97. Buckley, J. A., French, J. B., and Reid, N. M., Proc. New Concepts Symp. and Workshop on Detection and Identification of Explosives, Reston, VA, 1978, 109. NTIS No. PB 296055.
98. Davidson, W. R., Thomson, B. A., Akery, A. M., and Sleeman, R., Proc. First Int. Symp. on Explosive Detection Technology, Atlantic City, NJ, 1991, 653.
99. Neudorfl, P. and Elias, L., *J. Energ. Materials*, 4, 415, 1986.
100. Tanner, S. D., Davidson, W. R., and Fulford, J. E., Proc. Int. Symp. on Analysis and Detection of Explosives, Quantico, VA, 1983, 409.
101. Davidson, W. R., Stott, W. R., Akery, A. K., and Sleeman, R., Proc. First Int. Symp. on Explosive Detection Technology, Atlantic City, NJ, 1991, 663.
102. Sleeman, R., Bennett, G., Davidson, W. R., and Fisher, W., Proc. Contraband and Cargo Inspection Technology Int. Symp., Washington, D.C., 1992, 57.
103. McLuckey, S. A., Glish, G. L., Asano, K. G., and Grant, B. C., *Anal. Chem.*, 60, 2220, 1988.
104. Asano, K. G., McLuckey, S. A., and Glish, G. L., *Spectrosc. Int. J.*, 8, 191, 1990.
105. McLuckey, S. A., Glish, G. L., and Asano, K. G., *Anal. Chimica Acta*, 225, 25, 1989.
106. Conrad, F. J., Hannum, D. W., Grant, B. C., McLuckey, S. A., and McKnown, H. S., Proc. 3rd Int. Symp. on Analysis and Detection of Explosives, Mannheim-Neuostheim, Germany, 1989, 35–1.
107. Stafford, G. C., Kelley, P. E., Syka, J. E. P., Reynolds, W. E., and Todd, J. F. J., *Int. J. Mass Spectrom. Ion Processes*, 60, 85, 1984.
108. McLuckey, S. A. and Glish, G. L., Proc. 35th ASMS Conference on Mass Spectrometry and Allied Topics, Denver, CO, 1987, 771.
109. Eckenrode, B. A., Glish, G. L., and McLuckey, S. A., *Int. J. Mass Spectrom. Ion Processes*, 99, 151, 1990.
110. McLuckey, S. A., Asano, K. G., and Glish, G. L., Proc. 36th ASMS Conference on Mass Spectrometry and Allied Topics, San Francisco, CA, 1988, 1108.
111. McLuckey, S. A., Goeringer, D. E., and Glish, G. L., *Anal. Chem.*, 64, 1434, 1992.
112. Alcaraz, A., Martin, W., and Andresen, B. D., Proc. 39th ASMS Conference on Mass Spectrometry and Allied Topics, Nashville, TN, 1991, 158.
113. Fine, D. H., Yu, W. C., Goff, E. U., Bender, E. C., and Reutter, D. J., *J. Forensic Sci.*, 29, 732, 1984.
114. Jackson, R. and Bromberg, E. A., Proc. 3rd Int. Symp. on Analysis and Detection of Explosives, Mannheim-Neuostheim, Germany, 1989, 42–1.
115. Grant, B. C., Goering, D. E., Hart, K. J., McLuckey, S. A., and Glish, G. L., Proc. 39th ASMS Conference on Mass Spectrometry and Allied Topics, Nashville, TN, 1991, 166.
116. Glish, G. L., McLuckey, S. A., Grant, B. C., and McKown, H. S., Proc. First Int. Symp. on Explosive Detection Technology, Atlantic City, NJ, 1991, 642.

Chapter 7

Use of Isotope Ratios in Forensic Analysis

Jean Louis Brazier

CONTENTS

0-8493-8252-1/95/$0.00+$.50

I. INTRODUCTION

The fundamental basis of chemistry is not the chemical element itself, but more precisely, the isotope. So that precise information on the intimate origin of a chemical compound consists not only of a perfect description of the whole and global population of molecules but more accurately of a scrupulous sorting of the various isotopomers which constitute this population. Starting from raw materials with various isotopic compositions and undergoing various isotope effects all along the synthetic or biosynthetic pathways, molecules progressively acquire a characteristic isotopic composition that constitutes their own "isotopic signature". It is obvious that a complete and meticulous sorting of each isotopomer would be quite unrealistic, but the measurement of isotope ratios of each of its constituting elements can act as the specific isotopic signature. Table 1 gathers the natural stable isotopes of interest in organic chemistry and their natural abundance. This isotopic signature can be self-generated during synthetic processes without any intentional modification. This signature can also be intentionally produced by adding precursors artificially enriched with stable isotopes in the reactional medium where the synthesis or biosynthesis takes place. Therefore, this intentionally modified isotope composition can be considered as an isotopic signature of the commercial or industrial property of the compound.

Consequently, any organic molecule displays a different specific isotopic signature according to its natural or synthetic origin. If this molecule is of natural origin it can bear the signature of the vegetal or animal kingdom from which it had been generated. When a molecule has been

TABLE 1.

List of Stable Isotopes of Hydrogen, Carbon, Nitrogen, Oxygen, and Sulfur, with their Atomic Weights and Natural Relative Abundances (%)

Element	Atomic weight	Natural relative abundance (%)
^1H	1.0078	99.985
^2H	2.0141	0.015
^{12}C	12	98.89
^{13}C	13.003	1.11
^{14}N	14.003	99.63
^{15}N	15.000	0.37
^{16}O	15.994	99.759
^{17}O	16.999	0.037
^{18}O	17.999	0.204
^{32}S	31.972	95.05
^{34}S	33.967	4.22

synthesized by a plant, it bears the signature of the botanical species because of metabolic differences, isotope effects, and isotopic discrimination occurring during the photosynthetic process (C3 plants, C4 plants, C3-C4 intermediates, CAM). The knowledge of the origin can be more precise and give information on the various factors which slightly modify the photosynthetic metabolism. Thus, environmental factors such as altitude, climate, temperature, hygroscopy, pluviometry, and the isotopic composition of both local CO_2 and water influence isotope ratios. With sufficient analytical resolution, an approach on the geographic and seasonal locations can be obtained. By determining the values of isotope ratios of carbon which correspond to the CO_2 source and of hydrogen and oxygen which correspond to water, it is possible to locate the origin of a natural compound both in time and space. This kind of information can be very useful in forensic analysis.

If the molecule is of synthetic origin it can bear the signature of its origin according to the isotopic composition of the chemical precursors, solvents, catalysts, and potential impurities. The accurate determination of the isotope ratios of such a molecule allows it to authenticate its origin and can be used as a legal proof of commercial and industrial property. A chemical compound can be isotopically signed, batch to batch, by adding different amounts of one or several artificially labeled precursors during the synthesis. The isotopic enrichment of each batch being different, each batch holds its own isotopic signature. Figure 1 shows how the various levels of information can be given by the measurement of isotope ratios in order to provide the isotopic signature of an organic compound.

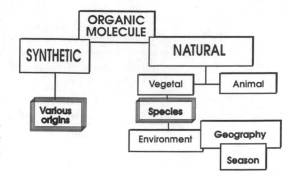

Figure 1.
The various levels of information given by the measurement of isotope ratios in order to obtain the "isotopic signature".

II. SOURCES OF VARIATIONS OF THE ISOTOPIC COMPOSITION

A. General Aspects

The major carbon reservoirs on earth, which are the source of either synthetic or natural organic molecules, display a wide range of $\delta^{13}C$ values from 0 $\delta‰$ for marine carbon to –50 $\delta‰$ for natural gas. A steady-state condition presently exists among all the isotopic compositions of the various terrestrial carbon reservoirs. The isotopic composition of atmospheric CO_2 (which is the raw material for photosynthetic processes) varies from –7 $\delta‰$ for oceanic air to $\delta^{13}C$ values of about –10 $\delta‰$ over rural areas. Isotopic variations can be induced when the proportion of CO_2 from the combustion of fossil fuels varies according to the location and season.

For terrestrial plants $\delta^{13}C$ values found vary from –8 $\delta‰$ to about –35 $\delta‰$ according to the type of photosynthetic mechanism. In C3 plants the $\delta^{13}C$ range is about –22 to –35 $\delta‰$; in C4 plants the range is about –8 to –20 $\delta‰$. The values of $\delta^{13}C$ for terrestrial animals range between –15 and –25 $\delta‰$. Modulations of these values are due to the isotopic variations all along the food chain. Petroleum and coals which give most of the precursors for organic chemistry and industrial synthesis display $\delta^{13}C$ values that range between –25 and –30 $\delta‰$; some of them are in the range –18 to –33 $\delta‰$. Last, the most negative $\delta^{13}C$ values are found in natural gas; they range between about –30 $\delta‰$ and –50 $\delta‰$. The variations of isotopic abundance of these precursors are the first arguments for isotopic authentication.

B. Sources of Isotopic Variations in Natural Compounds from Plants

The fractionation of carbon isotopes that occurs during photosynthesis can be understood in terms of very small differences in the physical and physicochemical properties between ^{12}C and ^{13}C. The lighter isotope diffuses and reacts more rapidly. So variations among plants can be under-stood in terms of differing internal partial pressures under different conditions. The basis of isotope fractionation has been reviewed by O'Leary et al. and some fundamental rules exposed.[1]

- The heavier isotope concentrates in the more constrained environment (in terms of atom bonding)
- The heavier isotope is transformed more slowly
- Fractionations in chemical processes are generally larger than those in physical processes
- Fractionations with enzymes are often smaller than those for the corresponding chemical reactions
- Changes in temperature and other reaction conditions may cause large changes in the isotopic fractionation in enzymatic reactions
- Terrestrial C3 plants show a large isotope fractionation
- C4 plants show a small isotope fractionation
- Isotope fractionation varies with the internal CO_2 concentration

The depletion in ^{13}C of the biomass can be attributed to the step of carbon assimilation by plants. In the principal mechanism, atmospheric CO_2 is directly introduced into the Calvin cycle by the ribulose 1,5 biphosphate carboxylase. This cycle is termed "C3 Cycle" because the carboxylation product is phosphoglycerate, a molecule with a 3-carbon skeleton. A fractionation by isotope effect occurs during this carboxylation step which is responsible for the ^{13}C depletion observed in the C3 plants which are the majority of vegetal species. Therefore, most of the plants cultivated for food supply are C3 plants, with two major exceptions: corn and cane (Figure 2).

In plants possessing the C4 photosynthetic pathway, CO_2 has to pass through a C4 cycle (intermediate molecules with four carbon atoms) before entering the Calvin cycle. The C4 cycle involves carboxylation of phosphoenolpyruvate (PEP) in the mesophyll cells, transfer of the C4 compounds to the bundle sheath cells, followed by a decarboxylation of these compounds. Isotopic discrimination in C4 plants depends on the fractionation occurring during CO_2 diffusion into the leaf and fractionation arising during the first carboxylation step. The CO_2 released by the decarboxylation of C4 compounds would be fixed by RuBP carboxylase (Rubisco) with no fractionation. In reality, as shown by O'Leary, CO_2 can diffuse outward from the bundle sheath cells which allows Rubisco to express some of its fractionation (Figure 3).[2] Environmental variations in each carbon isotope discrimination have been studied in C4 plants, but generally the $\delta^{13}C$ in C4 plants is less sensitive to changes in environmental conditions compared to $\delta^{13}C$ in C3 plants. The main environmental factors

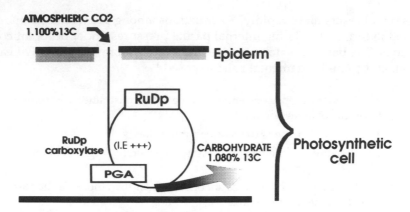

Figure 2.
The C3 cycle of photosynthetic pathways: Calvin cycle. RuDp = ribulose diphosphate, PGA = phosphoglyceraldehyde, and I.E. = isotope effect.

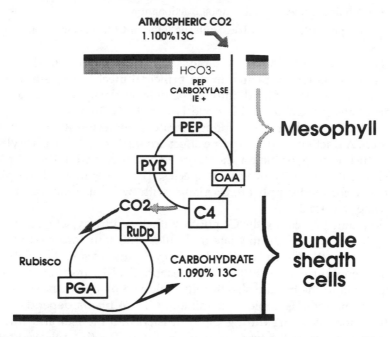

Figure 3.
The C4 photosynthetic pathways: Hatch-Shalck cycle. PEP = phosphoenolpyruvate, PYR = pyruvate, OAA = oxaloacetic acid, RuDp = ribulose diphosphate, PGA = phosphoglyceraldehyde, and I.E. = isotope effect.

which can slightly modify carbon isotope discrimination are temperature, CO_2 concentration, and air humidity.[3] Seasonal changes in carbon isotope discrimination of C4 plants were reported first in 1978.[4] For example, in *Zea mays* $\delta^{13}C$ values of the dry matter and soluble sugars were found to become less negative with decreasing leaf age.[5] In recent years, higher plants with C3-C4 intermediate photosynthetic pathways have been described.[6]

Generally, carbon isotope discrimination in C3-C4 intermediates is C3-like because only a small fraction of the total carbon fixed is fixed in the bundle sheath.[7]

Crassulacean acid metabolism (CAM) can be considered to be the temporal coupling of C3 and C4 assimilation and metabolic transfer within cells. CAM provides an adaptation acting as a survival mechanism in extreme habitats. Discrimination in C3 plants is usually fixed within a relatively narrow range of 3 to 4 δ‰, reflecting the balance between diffusion and carboxylation limitations expressed by Rubisco.[8] For C4 plants, the low discrimination products of PEPc are tempered by CO_2 leakage from the bundle sheath allowing discrimination by Rubisco to be expressed. These two processes combined to make up CAM account for a carbon isotope discrimination between 2 and 22 δ‰ (equivalent to $\delta^{13}C$ values of −10 to −30 δ‰). Besides CO_2, the other important material for biosynthesis of organic molecules by plants is water. The isotopic composition of hydrogen and oxygen incorporated in plant carbohydrates greatly depends on the isotopic composition of precipitation water.

As early as 1954, Dansgaard demonstrated that the variations of the $^{18}O/^{16}O$ and $^2H/^1H$ ratios were directly correlated with mean annual temperature.[9] The temperature of condensation of water has a strong influence on its isotopic composition. In most cases, the isotopic composition of groundwater is quite similar to that of precipitation.[10] In 1977, Yapp et al. showed a direct correlation between geographic and temporal variations in the isotopic composition of plant cellulose and variations in mean annual temperature.[11] So, the use of oxygen and hydrogen isotope ratios can provide excellent indicators of environmental parameters.

It is important to remember that the isotopic composition of water available in the environment and which can be taken up by plants is variable, and these variations are described as "climatic effects" by Ziegler et al.[12] There are marked differences between the isotopic composition of precipitations falling in summer and winter, as water or snow, and across latitudinal and elevational gradients. The measurement of deuterium concentration in sugars from Israelian orange juice shows a slight increase with the season. Until the end of January, $\delta(^2H)$ of nitrate esters of sugars are between about −30 and −20‰, but from February until the end of May, the $\delta(^2H)$ variation ranges between −20 to −10‰.[13] Because water molecules having the lighter isotope have a higher vapor pressure, they are preferentially lost during evaporation from the leaves where an enrichment of heavy isotopes takes place.

For example, the analysis of the nitrate esters of sugars carried out from beet sugar of North America shows a gradient of deuterium concentration with latitude from $\delta(^2H) = -160\text{‰}$ for Canada, to $\delta(^2H) = -122$ and -127‰ for Michigan and Illinois, down to $\delta(^2H) = -109\text{‰}$ in Texas.[13] In addition, the isotopic composition of water may be influenced, inside the leaf, by metabolic processes that produce or consume both oxygen and hydrogen.

Thus, the precise determination of hydrogen and oxygen isotope ratios in organic molecules biosynthesized by plants can afford important information on the environmental factors where the plant grows.

III. THE MEASUREMENT OF STABLE ISOTOPE NATURAL ABUNDANCE VARIATIONS

As the isotopic composition of an individual molecule can constitute the "isotopic signature" of the origin of the compound, authentication of the origin of a product, either for scientific or forensic purpose, necessitates very precise isotopic analysis. The magnitude of the natural abundance is small (in the order of one part of 10^4 of the major isotope). So, the analytical methodologies which can be used for such isotopic determinations must be able to resolve changes in the order of one part in a million. A wide variety of techniques have been used to measure stable isotope ratios. They are based on physical specific properties of the isotopes: mass spectrometry using the differences in atomic weights, nuclear magnetic resonance (NMR) using the resonance properties of several isotopes, and atomic emission using the specific emission lines generated by the various isotopes of the same element. Deuterium NMR at the natural abundance can be used for the investigation of site-specific natural isotope fractionation (SNIF-NMR).[14] This method has been extended to the ^{13}C spectroscopy.[15] The main advantage of NMR is the ability of measuring the isotopic enrichment on specific molecular sites. At the present time, NMR necessitates rather large amounts of a perfectly well-purified molecule in order to carry out the isotopic measurement. Moreover, the technique cannot be coupled to a separative method for the analysis of complex mixtures.

Detection by atomic emission is able to measure isotope ratios of the various elements constituting organic molecules: C, H, N, and O. This mode of element or isotope selective detection can be coupled with gas chromatography (GC-AED) but, presently, the detection of isotope ratios is not sensitive enough to detect variations within the range of natural isotopic composition. Nevertheless, this new methodology is quite a powerful tool for the isotopic analysis of molecules artificially enriched with stable isotopes.[16,17]

A. Isotope Ratio Mass Spectrometry

Only mass spectrometry is sufficiently precise to allow the measurement of variations between the natural isotopic abundances of small samples. Isotope ratio mass spectrometry (IRMS) which was developed for the analysis of light gas stable isotopes has been extensively improved during the last 5 years. It allows the determination of isotope ratios with a very high level of precision and accuracy from very small amounts (nanograms) of material and can be coupled to gas chromatography. So, GC-C-IRMS (gas chromatography-combustion-isotope ratio mass spectrometry) has been developed for the determination of isotope ratio profiles from complex mixtures and can be considered now as a very powerful tool in forensic analysis.

Isotope ratio mass spectrometers are designed for the continuous analysis of adjacent ion beams which necessitates a very high stability. As in organic mass spectrometry, gas introduced into the ionization source via a gas inlet system is ionized by electron impact and the resulting ions accelerated into a magnetic field. The magnetic field strength is usually kept constant as a permanent magnet or an electromagnet with fixed current. The flight tube forms an arc of circle and passes between the poles of the magnet. As the ion beam travels down the tube, it is separated into beams of different radii corresponding to different masses. A particular radius, and hence mass, is selected by a slit at either end of the flight tube. Then an optimized tuning of the accelerating voltage allows continuous collection of the ion currents of interest. Ion beams are generally collected by an array of Faraday's cups, one for each ion beam. The ion current from the cup is proportional to the number of incident ions and hence to the partial pressure of the corresponding isotopic molecular species in the sample gas.

The most noticeable difference between analytical isotope ratio instruments lies in the peak shape observed by scanning the magnetic or electric fields. In analytical work by organic mass spectrometry, a mass range is scanned to obtain a spectrum of mass peaks which are characteristic of the chemical composition. In IRMS, the chemical composition of the sample is known and the fields are held constant so that the variations of isotopes in one chemical species may be measured with high precision. Thus, an analytical organic mass spectrometer requires a very narrow peak to distinguish closely spaced masses, whereas in IRMS, very broad peaks are required for high stability in the amplitude measurements of isotopic ratios.

1. Carbon Isotopes

$^{13}C/^{12}C$ ratios are determined on a monocarbon molecule: the CO_2 gas. So, organic carbon has to be converted to CO_2 by combustion over CuO and cryogenically purified. CO_2 can be then transferred into evacuated sample tubes for sequential IRMS analysis or immediately directed to the spectrometer by a transfer line. This last device is a continuous flow IRMS (CF-IRMS). The statistical combination of the isotopes of carbon (^{12}C and ^{13}C) and oxygen (^{16}O, ^{17}O, ^{18}O) to generate the CO_2 molecules gives rise to the formation of various isotopomers whose molecular weights are, respectively, 44, 45, and 46. Thus, for measuring carbon isotope ratios, three ion beams are generated and recorded in the IRMS; they correspond to the masses of the various isotopomers of CO_2. In order to obtain a high precision and a high accuracy, reference gases of known isotopic composition are used and a dual inlet system allows an alternative admission of both sample gas and reference gas into the ionization source via a gas-switching valve. The measurement of the various ion beams allows the calculation of the ^{13}C enrichment of the sample: $\delta^{13}C(\text{‰})$. The ^{13}C abundance is generally expressed as $\delta^{13}C(\text{‰})$ according to the following relation:

$$\delta^{13}C(\text{‰}) = ([(^{13}C/^{12}C) \text{ sample}/(^{13}C/^{12}C) \text{ PDB}] - 1) \times 1000$$

This delta $\delta^{13}C(\text{‰})$ value measures the variations in parts per thousand of the carbon isotope ratio from the standard. For carbon, PDB was selected as the international reference.[18] PDB is the Pee Dee Belemnitella (a fossil from the Pee Dee geological formation of South Carolina). The $^{13}C/^{12}C$ ratio from the calcium carbonate of this fossil is 0.011237. Compared to PDB, most of the natural compounds display a negative delta value. The same notation is also used for $^2H/^1H$, $^{15}N/^{14}N$, $^{18}O/^{16}O$ ratios.

In the equation above, $^{13}C/^{12}C$ only refers to the following isotopomers: $^{13}C^{16}O^{16}O$ and $^{12}C^{16}O^{16}O$. So, in addition to corrections corresponding to the overlapping of the very much larger 44 mass on the smaller mass 45 beam and to the "zero enrichment" which corresponds to very small differences in mass fractionation that may occur in the capillary crimps, the "Craig" correction for oxygen has to be performed. This correction is necessary because of a difference in ^{17}O composition between the sample CO_2 and the reference CO_2. The correction for the ^{17}O effects uses the measurement of the mass 46/mass 44 ratio which gives the ^{18}O difference between sample and reference. This ^{17}O correction must be applied on each sample analyzed. For example, the international CO_2 standard PDB has a contribution of approximately 6% at mass 45 due to ^{17}O and 0.2% of mass 46 derived from isotopic species containing ^{13}C and ^{17}O but not ^{18}O. So, when using a triple collector instrument which simultaneously records these ions at m/z 44, 45, 46, the correction formulas are

$$\delta^{13}C = 1.0676 \text{ delta } (45/44) - 0.0338 \text{ delta } {}^{18}O$$

and

$$\delta^{18}O = 1.0010 \text{ delta } (46/44) - 0.0021 \text{ delta } {}^{13}C$$

2. Nitrogen Isotopes

Nitrogen isotopes can be determined after quantitative oxidation to minimize CO and hydrocarbon production and careful reduction to convert nitrogen oxides into molecular N_2. A nitrogen analyzer can be interfaced to an IRMS and be used as a continuous flow device.[19] When analyzing nitrogen, a problem can arise owing to traces of nitrous oxide (NO) causing an interfering peak at mass 30. A correction has to be performed when ^{15}N enrichments are low (less than 5%).

3. Hydrogen Isotopes

The combustion of organic samples produces CO_2 and also generates water from organic hydrogen. This water formed during combustion can be reduced into H_2 by passing over uranium turnings held at 800°C. It is then possible to perform the $^2H/^1H$ ratio measurement from the resulting molecular H_2. The deuterium to hydrogen ratio is measured by the intensity ratio of H_2^+ and HD^+ peaks at masses 2 and 3, respectively. However, there is also an interfering peak at mass 3 due to $H3^+$ which can be formed by collisions in the ion source.

4. Oxygen Isotopes

Sample preparation for ^{18}O analysis involves an exchange of the ^{18}O content of water with CO_2 followed by the IRMS determination of CO_2.[20] As for carbon, hydrogen, and oxygen, isotope ratios can be expressed as $\delta‰$ using an internal standard which is SMOW (Standard Mean Ocean Water).

B. Continuous Flow-Isotope Ratio Mass Spectrometry

For forensic purposes, it is often of great importance to authenticate samples which are complex mixtures such as flavors, drugs of abuse, food, beverages, etc. The criteria for authentication can be isotopic profiles of the mixture, i.e., measurement and recording of isotope ratios of various chemical compounds which are the components of the sample. This kind of determination needs a high resolution separative technique to be coupled with IRMS. So, a new form of continuous flow IRMS termed GC-Combustion-IRMS has been developed. In this technique a high resolution capil-

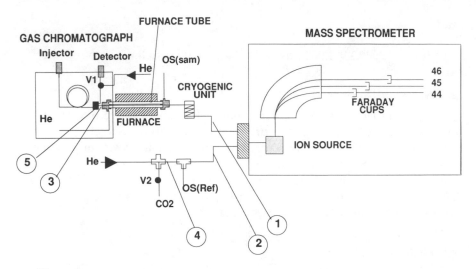

Figure 4.
Scheme of gas chromatograph-combustion-isotope ratio mass spectrometer (GC-C-IRMS): 1, sample inlet capillary; 2, reference inlet capillary; 3, sample line; 4, reference line; 5, splitter union; and OS = open split.

lary gas chromatographic column allows the high resolution separation of organic compounds which constitutes the mixture under analysis. This chromatographic system is located before a capillary combustion oven where the compounds corresponding to the chromatographic peaks are combusted and their organic carbon transformed into CO_2. This oven can be a capillary catalytic oven (0.5 mm inside diameter) filled up with CuO and maintained at 840°C.[21] Figure 4 displays the scheme of a GC-combustion-IRMS.

One of the original concepts in IRMS is the use of a dual inlet for alternate admission of a sample and a reference gas and subsequent comparative isotope ratio measurements. In the conventional dual inlet system the sample and reference gases are expanded in variable volume chambers and a change-over valve allows the alternate admission of the sample and reference gases into the ion source. The entire inlet is carefully designed so that isotopic fractionation does not occur. A new continuous flow dual inlet has been designed. This dynamic interface consists of two parallel gas lines: a reference inlet line and a sample inlet line. These lines draw the sample and reference gas to a switching valve via two open splits. This valve ensures the admission of both gases into the ion source of the spectrometer. On the reference side of the dynamic interface, a pressure-regulated flow of helium continuously sweeps the capillary where pulses of reference CO_2 can be injected. The open splits completely isolate the mass spectrometer from any pressure transients which may occur during valve switching in the interface system and ensure a high stability

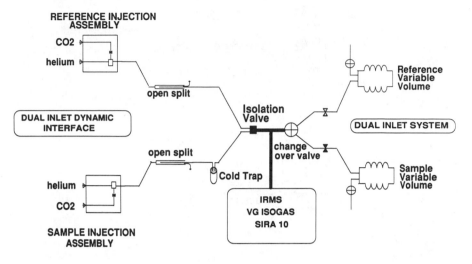

Figure 5.
Continuous flow-isotope ratio mass spectrometer (CF-IRMS). Comparative schemes of the classical dual inlet system and of the dual inlet dynamic interface.

in the ion source. A high efficiency cryogenic water trap ($-100 \pm 1°C$) is fitted along the sample inlet line. Figure 5 displays the comparative schemes of the classical dual inlet system and of the dual inlet dynamic interface which allows the on-line preparation of the CO_2 to be isotopically analyzed. Obviously, water must be prevented from entering the mass spectrometer because the partial protonation of CO_2 induces the formation of CO_2H^+ ions, thus interfering with the isotopic measurement. After the switching over valve, CO_2 enters the ion source and ions at m/z 44, 45, and 46 are continuously recorded. With such a device it is possible either to measure the $^{13}C/^{12}C$ ratio of all the chromatographic peaks eluted from the column and to obtain the carbon isotopic profile of the sample or to select only one or several compounds of interest using a system of heart-cut. Good precision has been demonstrated for ^{13}C analysis of small samples by such a GC-C-IRMS: 0.14‰.[22]

The development of multidimensional gas chromatography allows the design of on-line coupling of a multidimensional GC (MDGC) system to an isotope ratio mass spectrometer.[23] In such devices, the chromatographic stage of the MDGC-IRMS consists of two columns which can combine various and complementary properties: packed-capillary, capillary-capillary, megabore-microbore, achiral-chiral, etc. By means of a valveless switching device, only preseparated peaks or groups of peaks are quantitatively transferred from one column to the other, only using differential pressures. Coupling such a multidimensional device to IRMS allows solution to difficult problems of authentication from very complex mixtures like aromas and affords more specific information. For example,

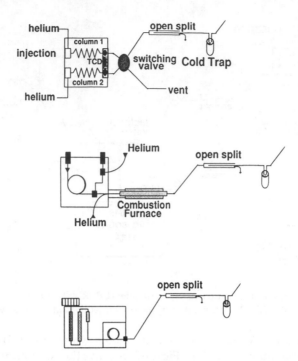

Figure 6.
The various sample preparation devices which can constitute the sample line of the dual inlet dynamic interface: (A) Gas chromatograph for CO_2 separation from mixtures of gases such as expired gas; (B) GC-Combustion-IRMS; and (C) Elemental analyzer.

both ratios of the enantiomeric distribution and isotopic ratios can be obtained simultaneously during the same chromatographic run of an essential oil. In the same way, minor components which can be specific markers of the authenticity can be more easily separated from major constituents and isotopically measured.

Besides analytical systems, in which gas chromatography is coupled with IRMS like GC-IRMS, GC-C-IRMS, and MDGC-C-IRMS, it is possible to interface IRMS with other devices allowing the preparation of gas samples via a dual dynamic interface. Elemental analyzers can be directly coupled on the sample line of a dynamic dual inlet and allowed to process pure solid or liquid samples or fractions collected after HPLC separation. It is also possible to associate micro-CO_2 generators that can generate microvolumes of CO_2 using either chemical or enzymatic reaction from substrates to be analyzed for their isotopic content. The resulting CO_2 produced by the reaction is on-line purified by gas chromatography and immediately measured for its ^{13}C content (Figure 6).

Such a system which has been developed by R. Guilluy et al. is able to measure a difference of 0.2‰ between baseline and enriched samples from 10 nmol of CO_2.[24,25] In conclusion, IRMS alone or coupled with highly resolutive chromatographic systems and/or with devices allowing an accurate preparation of the gas samples to be analyzed affords a very powerful tool for isotopic measurements from small samples and consequently can provide useful information in the field of forensic authentication.

IV. THE USE OF ISOTOPE RATIOS FOR AUTHENTICATION AND FORENSIC ANALYSIS

We have seen that owing to small differences in physical properties, the various isotopes of the chemical elements that constitute the matter display slight variations in their behavior during chemical and biochemical reactions. Series of tiny isotope effects progressively record inside the molecule the isotopic "memory of its origin". This memory is its isotopic composition which reflects the isotopomeric composition and acts as a true isotopic signature. The recent developments in IRMS allow very precise and accurate determinations of isotope ratios from very small samples. So these measurements of isotope ratios can afford objective elements and arguments suitable to courts of laws and justice. The main fields of interest are

- Adulteration of regulated products like food, beverages, aromas, essential oils, etc.
- Frauds
- Research of the origin of drugs of abuse — authentication of the origin
- Protection of patents for synthetic materials or commercial mixtures of chemicals
- Research of the natural or synthetic origin of biomolecules fraudulently used to dope sportsmen or sport animals (horses, dogs, etc.)
- Intentional labeling with small amounts of stable isotopes to obtain a personal isotopic signature in order to protect an industrial property

A. Adulteration of Regulated Products

Adulteration of a product consists in making it impure by fraudulous addition of a foreign or inferior substance. The result is either an alteration of the product and of its quality or a falsification. The falsification is a voluntary act with the intention of misabuse. The falsification may be more or less sophisticated and its sophistication as well as its cost increases with the improvement of analytical methods, and there is always a competitive technical race between the falsificator and the researcher in frauds. When sophisticated methods are used, the analytical procedures using IRMS alone or coupled with chromatographic systems offer a potent tool to forensic analysis. In this area, the example of vanillin is exemplary.

1. Vanillin

Natural vanillin is derived from vanilla beans produced by *Vanilla planifolia* or other species of Vanilla (*tahitensis* in Tahiti, for example). Vanilla is a plant using CAM photosynthetic pathways. Vanilla flavor is quantitatively very important and its production constitutes a multimillion dollar-a-year industry. The cost of 1 kg of natural vanilla extract is presently negotiated at about $2500. Synthetic vanillin, nature-identical, can be synthesized from lignin derived from wood pulp; it can be also prepared from guaiacol (from coal tar) or from eugenol (from clove oil). This low-price vanillin is used for the fraudulent adulteration of natural vanillin. The cost of 1 kg of synthetic vanillin is about $10. So, adulteration of vanilla extracts with synthetic lignin-derived vanilla is a major problem. Considerable effort has been directed toward the development of reliable and accurate methods for the detection of such an adulteration.

IRMS has proved to be a method of choice for the detection of this fraud.[26-28] Table 2 shows that a significant difference exists between the $\delta^{13}C$ total values of natural and synthetic vanillin when the whole molecule is combusted to transform all the carbon atoms into CO_2.

As synthetic vanillin shows a $\delta^{13}C$ more negative than natural vanillin, it is obvious that falsificators tried to artificially enrich "synthetic vanillin" in order to obtain a falsified "natural" vanillin in which the $\delta^{13}C$ could be identical to that of true natural vanillin. The simplest way to perform such an adulteration consists in labeling the methyl group of vanillin with ^{13}C. As a consequence, in order to find the [methyl-^{13}C] falsified vanillin, the molecule has to be demethylated. The demethylation can be carried out with hydriodic acid which transforms vanillin into dihydroxy benzaldehyde (DHB) and allows to collect the [methyl-^{13}C] group as [^{13}C-methyl iodide]. The measurement of the carbon isotope ratio can be performed on

TABLE 2.

^{13}C Enrichment of Various Vanillin Molecules $\delta^{13}C$‰ PDB

Source of vanillin	$\delta^{13}C$ total
Vanilla (Madagascar)	−20.4
Vanilla (Java)	−18.7
Lignin	−27.0
Eugenol	−30.8
Guaiacol	−32.7

TABLE 3.

^{13}C Enrichment of Various Vanillin Molecules. δ^{13}C‰ PDB, Natural, Semisynthetic from Lignin, ^{13}C Enriched Vanillin from Lignin (Comparison of the Whole Molecule, the Dihydroxybenzaldehyde Moiety, and the Methyl Group)

Sample	δ^{13}C total	δ^{13}C DHB	δ^{13}C CH$_3$
Vanilla (Madagascar)	−21.4		−24.0
Lignin	−27.3		−28.4
Lignin	−27.0	−26.7	
^{13}C enriched lignin	−20.6		+25.8
^{13}C enriched lignin	−20.0	−26.7	

From Krueger, D. A. and Krueger, H. W., *J. Agric. Food Chem.*, 31, 1265, 1983. With permission.

both moieties of the molecule. Results from Krueger et al. are shown in Table 3.[28]

Natural vanilla samples from various geographic sites have been analyzed by IRMS. It was shown that the methyl carbon is generally 3 to 5‰ more negative except in Tahitian samples (from *Vanilla tahitensis*). This isotopic composition is that expected from a C4 photosynthetic pathway (−10‰), while other vanillin samples have isotopic composition expected from the Calvin C3 cycle (−25‰). In fraudulent [^{13}C-methyl] vanillin, the methyl carbon is obviously largely positive.

In 1985 Krueger et al. described a new method to detect lignin-vanillin labeled with ^{13}C on the carbonyl group of the aldehyde function.[29] In this method, vanillin is oxidized into vanillic acid (transformation of the aldehyde group into a carboxylic group) which is reacted with bromine and acetic acid. 3,5-Dibromo-4-hydroxybenzoic acid is formed which is decarboxylated as CO_2 with an excellent yield and negligible isotope fractionation. The CO_2 evolved is measured for its ^{13}C contents by IRMS. A positive enrichment of the carbonyl carbon atom points out the adulteration. Some results from D. Krueger are shown in Table 4.[29]

TABLE 4.

^{13}C Enrichment of the Whole Vanillin Molecule and the Carbonyl Group from Natural Vanillin, Semisynthetic, and ^{13}C Enriched Lignin-Vanillin

Sample	δ^{13}C total	δ^{13}C carbonyl
Bourbon vanilla	−21.4	−25.7
Lignin	−27.22 ± 0.02	−37.3
^{13}C Enriched Lignin vanillin	−19.94 ± 0.03	+17.1

In various countries the official methodology for the control of vanillin includes the determination of carbon isotope ratios. This example with vanillin provides an exemplary illustration of the possible sophistication of adulteration and of the power of isotope ratio determination by IRMS. It shows that it is not only possible to measure the difference between natural vanillin and "synthetic" as well as isotopically falsified "natural" vanillin, but also to distinguish vanillin from different geographical origins or vanilla species (*V. tahitensis*).

2. Honey and Maple Syrup

In the mid-1970s a new sweetener, high fructose corn syrup (HFCS), became available because of the development of new techniques of production. This HFCS was rapidly used as a source of adulteration owing to its low price, even less than that of cane sugar. HFCS is a liquid sugar that could be easily added to a large variety of foods like honey, maple syrup, fruit juices, etc.

It was shown that honey was normally derived from flowering plants which are almost exclusively C3 plants.[30] HFCS is derived from corn which is a plant with a C4 photosynthetic metabolism. HFCS is produced by action of various enzymatic steps — alpha amylase, gluco-amylase, glucose isomerase — from corn starch. Honey being a very complex mixture of various kinds of polysaccharides, it is almost impossible to find a specific marker of its authenticity and measurements have to be carried out on the whole product. Table 5 from L. Doner shows the uniformity of $\delta^{13}C$ values of honeys from various plant families.[31]

The adulteration of honey by addition of HFCS results in elevated $\delta^{13}C$ values which can be determined by IRMS. The $\delta^{13}C$ value of honey shows a coefficient of variation of only 3.86%.[32] It can be compared to the range of $\delta^{13}C$ values of carbohydrates according to their vegetal origin gathered in Table 6.

The $\delta^{13}C$ values gathered in this table clearly show that IRMS could be able to detect the presence of carbohydrates from C4 plants added to products from C3 plants like honey. As early as 1979, the Association of Official Analytical Chemists adopted IRMS as the official method for detection of honey adulteration by HFCS. This official method set a conservative limit of $\delta^{13}C = -21.5‰$ as a proof of honey adulteration. Adulteration strategies shifted to the addition of inverted syrup from beet, which is a C3 plant; therefore, forensic analysis has to be improved by additional criteria.

As with honey, the same principles for adulteration and detection of fraud can be described for maple syrup. Maple syrup is obtained by concentration of the sap of sugar maple (a C3 tree) and is sold at a high price when marketed 100% pure. Here, too, the temptation is great to add

TABLE 5.

Values of $\delta^{13}C‰$ PDB of Honeys from
Various Botanical Families

Anacardiaceae	−25.0	Malvaceae	−24.7
Aquifoliaceae	−25.6	Onagraceae	−25.4
Compositae	−25.6	Palmae	−24.7
Cornaceae	−26.0	Polygonaceae	−24.6
Cyrillaceae	−24.2	Rosaceae	−26.1
Euphorbiaceae	−26.4	Rutaceae	−23.4
Labiatae	−25.5	Tamaricaceae	−25.1
Leguminosae	−25.3	Tiliaceae	−25.6
Magnoliaceae	−25.3		

From Doner, L., European Seminar on
Authentication and Quality Assessment of Food
Products, Martin, G., Ed., Nantes, 1991. With
permission.

TABLE 6.

Range of the $\delta^{13}C$ Values of Carbohydrates According to their
Photosynthetic Pathway

	C4 Plants $-9 < \delta^{13}C < -12‰$ PDB	C3 Plants $-24 < \delta^{13}C < -30‰$ PDB
Sucrose	Cane	Beet
		Maple
		Potato
Glucose + Fructose	Corn	Apple
		Honey
		Orange

cheaper substituents like cane sugar or HFCS for the adulteration of maple syrup. Because of the differences in the metabolic photosynthetic pathways of the sugars constituting these products, the measurement of isotope ratios is the unique tool able to detect the adulteration.

3. Fruit Juices

a. Orange Juice

Orange juice is a very important product worldwide. The variability of orange production on an annual basis and at various times throughout the year, as well as the increasing demand of the market, suggests that fraudulent actions are carried out to increase the orange juice supply by various kinds of dilutions or sweetener additions. A fruit juice can be obtained by a mechanical process from fruit; in the case of citrus fruit (especially oranges), the fruit juice will come from the endocarp of the fruit. A fruit juice can also be obtained from a concentrate. These concentrates are generally prepared on the site of production. The restoration of concentrate juices is made by adding water to the concentrate in order to obtain the same volume as the initial one. The control of a correct dilution of the concentrate can be performed by the measurement of isotopic ratios of water molecules (hydrogen and oxygen).

The matrix of orange juice is very complex, and besides adulteration by excessive dilution with tap water, other types of adulteration can be performed by addition of various sugars which is not indicated on the final product as requested by the food regulation: cane and beet sugars or HFCS. Addition of an excess of pulpwash solids or of other cheaper juices (grapefruit, etc.) can also be fraudulently performed. The simple dilution can be detected by checking the Brix. HPLC sugar profiles may be of interest in looking for cane or corn adulterants. Specific molecules which are markers of other fruits can sign the presence of an adulteration by juices of other fruits (naringin for grapefruit).

As citrus fruit comes from C3 plants, the measurement of carbon isotope ratios allows the detection of the presence of sugars coming from C4 plants (corn); the detection of beet sugar (C3 plant) is much more difficult, but the determination of the $^2H/^1H$ ratio of hydrogen atoms bound to the carbon atoms of sugar molecules allows the detection of such an adulteration. The isotopic composition of orange sugar ($\delta^{13}C$) and (δ^2H) of their nitrate esters allows the detection of the fraudulent addition of sugars from beet cane and corn.[33] In the method developed by Bricout, orange juice is diluted to 11.5° Brix and centrifuged. the supernatant fluid is separated from the pulp which is washed with acetone to extract the lipids and dried. The clarified supernatant is freeze dried. It is then combusted and the resulting CO_2 measured for its ^{13}C content. The determination of carbon isotope ratio is also performed on the dried pulp. The results obtained from these isotopic measurements show that the sugars of orange juice are in the range expected for C3 plants ($\delta^{13}C$ = −28.0‰ to −23.5‰). It was observed that the pulp has a slightly more negative value than sugar. The difference is generally less than 1.5‰. The addition of cane sugar (−11‰) or corn sugar (−13.5‰) will result in a final product with an increased $\delta^{13}C$ value and a greater difference in the $\delta^{13}C$ values between pulp and its supernatant. The mean $\delta^{13}C$‰ PDB value reported for the sugars from 40 orange juices is −25.1‰ with a standard deviation of 0.9‰. One can admit that at a value of −3σ or −22‰ it can be concluded that 20% corn or cane sugar has been added, but statistically at −3σ there is an above 2% chance of such a value occurring. The probability of a second sample to be at 3σ is $(0.02)^2$ and that of a third sample is $(0.02)^3$ = 0.0008%, which is a very small risk. Thus, multiple sampling and multiple sample processing can prove the adulteration.

Results by analysis of the nitrate esters of sugars are also very interesting. We have seen that an isotopic discrimination in the isotopes of hydrogen occurs according to the latitude owing to "climate effects". Cane sugar (δ^2H‰ SMOW = –41) and HFCS (δ^2H‰ SMOW = –31) show higher (δ^2H) values than beet sugar (δ^2H‰ SMOW = –100 to –160‰) and a similar value to orange juice (–22.10 ± 10, n = 40). Here the influence of the type of photosynthetic process, C3 or C4, and the climatic conditions at the geographical site of production combine themselves to offer an excellent means for detecting the adulteration of orange juice by addition of sugar from beet.

Isotopic analysis of water is also useful to test the authenticity of orange juices. Hydrogen and oxygen isotope ratios of ground water (which is tap water) are a function of environmental variables such as temperature and altitude. They always are negative and less than those of plants which grow using this water. So, alteration of orange juice by dilution, addition of orange pulp wash, or beet syrup prepared with this water can be detected by the measurement of the isotope ratio of water molecules. Adding beet inverted syrup to concentrates of orange juice causes δ^{18}O value to decrease. The more beet syrup is added the lower the resulting δ^{18}O value in the adulterated concentrate. After important collaborative studies, sufficient agreement has been demonstrated to recommend this isotopic method as official for the detection of beet syrup in frozen orange juice concentrates. If an orange juice is concentrated and then rediluted with tap water, the isotopic composition of the water of this reconstituted juice is very close to that of the water used; thus, deuterium and oxygen-18 analysis is an easy way to differentiate natural orange juice from reconstituted juices. This assumption has proved to be valid for other fruit juices (grapefruit, pineapple, apple).[34] To conclude on the control of orange juice, the enhancement of the sensitivity of adulteration detection should involve the determination of isotopic ratios of carbon, hydrogen, and oxygen from pulp and supernatant in order to gather all the information allowing assessment of the quality and the integrity of the product.

b. Apple Juice

The market of apple juice is very important and the same kinds of adulteration can be performed on this fruit juice. The δ^{13}C values for apple juice show more variation than do those of orange juice or honey. The mean value for all juice samples measured by Lee et al. was –24.2‰ and the isotope composition of subfractions (lipids, amino acids, proteins, polysaccharides, organic acids) ranged between –22.0‰ and –31.0‰.[35] This variability in δ^{13}C values is due to environmental differences which are influenced by the wide geographic distribution. The same isotopic methods can be used to detect corn and cane addition to apple juice.

Results of tests of apple-HFCS mixtures were presented to the AOAC for approval as an official method. This approval was granted in 1981 with a proof of adulteration limit of −20‰. The effect on the quality of apple juice was important. Since 1980 only an occasional juice or concentrate has been found that contains corn or cane sugar. Isotopic analysis remains the official technique used to maintain the integrity of the market of apple juices.

c. Grape Juice

Grapes are C3 plants and therefore the carbohydrates produced by them should have a $\delta^{13}C$ value of approximately −25‰. The same methods can be used for the isotopic analysis of grape juice. For the detection of adulterations in wines and the production of alcohol from foreign sugars fraudulently added to grapes, the main isotopic methodology is certainly SNIF-NMR developed by G. Martin. Exploiting the joint structural and quantitative dimensions of NMR spectroscopy, G. J. Martin and M. L. Martin have shown that deuterium is far from being randomly distributed in organic molecules. The large variations detected in the deuterium contents at the different molecular sites are related to specific kinetic or thermodynamic isotope effects, and the determination of site-specific natural isotope fractionation by nuclear magnetic resonance was proposed as a method of labeling without enrichment.[14] This method has proved to be particularly powerful in the field of wines which remains one of its privileged applications.[36–38]

B. Authentication of Geographical Origin — Tracing of Trafficking Routes

Isotopic discrimination induced by photosynthetic pathways and environmental factors records in every molecule the isotopic signature of its origin. The very high analytical qualities of IRMS applied to the elements which constitute organic molecules are able to read accurately this isotopic signature and to point out the origin of a product. Carbon isotope ratios give information on the botanical origin, but are only slightly modified by environmental factors. Hydrogen and oxygen isotope ratios bring a large part of environmental information and are very useful tools for the authentication of origin. Two important fields of applications can be found for the determination of the origin of molecules using isotope ratio mass spectrometry: the authentication of origin and the tracing of trafficking routes.

Some regions or countries want to give more commercial value to some of their products, generally a natural product with a label which authenticates its origin. This label of origin can be a specific isotopic

TABLE 7.

Values (‰) of Isotopic Enrichments (^{13}C, ^2H, ^{18}O) of Caffeine from Various Origins

Origin	Source	$\delta^{13}C$ PDB	δ^2H SMOW	$\delta^{18}O$ SMOW
Jamaica	Coffee	−28.8 ± 0.6	−132.5	+9.6
Kenya	Coffee	−29.8 ± 0.6	−135.5	+3.6
Brazil	Coffee	−28.2 ± 0.2	−157.3	+4.9
Sri Lanka	Tea	−31.7 ± 0.8	−223.6	+1.8
Darjeeling	Tea	−32.4 ± 0.6	−195.9	−4.3
China	Tea	−32.4 ± 0.6	−226.8	+1.2
Synthetic		−35.8 ± 0.2	−237.1	+13.0

signature. An example has been given by Dunbar with caffeine.[39] Caffeine is a trimethylxanthine found all over the world as a natural molecule, biosynthesized by various plants, and more specifically found in coffee beans and tea leaves. This molecule can also be prepared by chemical synthesis. Isotope ratios were determined for carbon, hydrogen, and oxygen from natural as well as synthetic caffeine molecules by IRMS, using the previously described techniques. The results are gathered in Table 7.

Isotope ratios of the three elements measured are quite the same for caffeine from Sri Lanka tea and China tea; therefore they cannot be differentiated by these criteria. In the coffee group, where the carbon isotope ratios are the same within the limits of error, the Brazilian $^2H/^1H$ ratio is distinguishable from the other two as is the Jamaican $^{18}O/^{16}O$ from the other two oxygen isotope ratios. The Darjeeling sample is separated from the others both by hydrogen and oxygen isotope ratios. This difference could be explained by the geographical specificities of the country which is mountainous and at a higher average altitude than most countries. Hydrogen isotope ratios fall into two groups which allow distinguishing caffeine from coffee and caffeine from tea.

It is desirable to determine the geographical source of illegal drugs and drugs of abuse such as morphine, heroin, cocaine, etc. An analytical strategy based on the measurement of isotope ratios could give useful information for tracing the trafficking routes. In 1979 Liu et al. tried to obtain isotopic information on the geographical origin of cannabis by measurements of carbon isotope ratios from both leaves and flowers of the plant.[40] In 1990 we began a program of research with the Department of Scientific and Technical Police in France, in order to locate the geographical origin of heroin samples. Pure heroin (diacetyl morphine) is produced in various countries. The drug is then mixed with a wide variety of chemical compounds. Such heroin mixtures, containing very different amounts of pure heroin, are then handled by dealers. In order to trace these drugs of abuse, it is of importance to obtain a profile of the "impurities" added to the drug and also to try to identify the geographical origin of heroin itself. The improvements of chromatography coupled to mass spectrometry, the power of computer-assisted processing of analytical data, and the development of on-line gas chromatography-isotope ratio mass spectrometry make it possible to adopt a dual plan for the determination of the impurities and the ^{13}C content of heroin samples, in order to trace the routes of distribution of this drug. First a GC-MS dedicated program is used for automatic or manual data processing in order to obtain chromatographic and mass spectral parameters for complex mixtures of impurities in heroin samples. This normalized data is then used to compare heroin samples from various origins. The "relative retention time program" which was developed and used is divided into three parts. The first part is a search for three reference compounds and validation of the reference retention times. The second part involves automatic processing of the whole mass-chromatogram by calculation of the relative retention time of any integrated peak and construction of a data matrix which is saved at the end of the process as a text file. The third part is interactive and enables specific portions of the chromatogram to be studied offering various options (control of peak purity, library search, comparison of spectra, etc.). The second part of the strategy which affords additional data is a carbon isotope analysis of heroin. After extraction of heroin and related compounds (residual alkaloids), the carbon isotopic profile is carried out using GC-combustion-IRMS. The determination of the carbon isotope ratio is performed after the extraction of only 20 mg of the mixture sample containing heroin. The advantage of using GC-C-IRMS is the possibility to measure the ^{13}C/^{12}C ratios of several peaks all along the chromatogram corresponding to other alkaloids from *papaver*, partially acetylated alkaloids, and added dilutants such as caffeine which is very often present in drug samples. This multiple isotopic information on a drug sample can be helpful in tracing the origin of the sample. This dual analytical strategy using both GC-MS for the determination of an "impurities profile" and GC-IRMS for the determination of an "isotopic profile" combines the power of these two mass spectrometric techniques for tracing trafficking routes.[41] On the basis of carbon isotope ratios and statistical analysis, the first results from some heroin samples are gathered in Table 8.

As heroin is a diacetylated derivative of morphine, more information can be afforded by the morphine moiety of the molecule, which can sign the geographical origin of the plant, and by the acetyl moiety, which can bear the signature of the illicit laboratory where the molecule is prepared (because of the various reagents and synthesis routes used to transform morphine into its diacetylated derivative).

TABLE 8.

Comparison of Heroin Samples of Various
Geographical Origins from Carbon Isotope
Ratios of the Whole Molecule

Comparison of origin	Statistical significance
Turkey vs. Niger	S
Turkey vs. Thailand	S
Turkey vs. Pakistan	S
Turkey vs. India	S
Niger vs. Thailand	NS
Niger vs. Pakistan	NS
Niger vs. India	S
Thailand vs. India	S
Thailand vs. Pakistan	NS
India vs. Pakistan	S

From Desage, M., Guilluy, R., Brazier, J. L., Chaudron,
H., Girard, J., Cherpin, H., and Jumeau, J., *Anal. Chim.
Acta,* 247, 249, 1991. With permission.

C. Authentication of the Natural or Synthetic Origin of Flavoring Substances

The regulating agencies of various countries (U.S., EEC, etc.) have set up various regulating frameworks concerning flavoring substances which can be used in foodstuffs. For example, the EEC frame-directive defines six categories of flavors: natural, nature-identical, artificial, smoke and process flavor, and flavoring preparations. Their definitions are mainly based on the process used for the production of flavors.

Natural flavors are obtained by physical processes including distillation and solvent extraction or enzymatic as well as microbiologic processes from a vegetal or animal raw material.

A nature-identical flavor is a flavor obtained by chemical synthesis or isolated by a chemical process. This substance must be chemically identical to a natural substance present in a vegetal or animal raw material.

Artificial flavors are chemically synthesized substances which are not chemically identical to substances naturally existing in vegetal or animal materials. Genetically engineered micro-organisms can be used to produce flavor for or during food elaboration processes. This approach is presently limited by both consumers and legislative acceptability due to risks involved in human consumption of such micro-organisms.

The market for flavoring substances for use in foodstuffs is extremely important and the craze for Nature increases the demand for natural substances. The world consumption of flavoring molecules by the industry of aromas and flavors was about 12,000 tons in 1990. A clear marketing advantage is brought by the label "natural". However, the cost of natural flavors is higher (×10) than the price of synthetic copies (nature-identical). So, adulteration and frauds can be important due to the difference between the respective costs of natural and synthetic molecules and the difficulty of differentiating between their origin. Natural flavoring molecules are produced from fruit juices, but owing to very small concentrations the part of active molecules afforded by fruit juices represents only 30 tons a year on the world market. Essential oils are an important source of natural active flavoring molecules: about 500 tons a year for citrus fruits, 400 tons from mint, and 600 tons from other essential oils. Other sources are oleoresins and molecules obtained by biotechnological processes.

It is clear that industry has long been interested in using isotopic analysis to authenticate flavor materials. In the U.S., FEMA (Federal Extract Manufacturers Association) recognized the importance of isotopic analysis and formed the Flavor Labeling Analytical Subcommittee. After several years, the group was granted separated committee status and renamed Isotopic Studies Committee (ISC). In the U.S., the labeling of flavors is regulated by the FDA.

In natural flavors from plants, isotope ratios of the various elements which constitute active flavoring molecules are determined mainly by the biochemical pathways of CO_2 and water in the plant cells. The sources of nature-identical flavors are petrochemically derived compounds which are the precursors used for chemical synthesis. The difference in isotope ratios (carbon, hydrogen) between organic molecules synthesized from fossil fuels and biosynthesized by plants can be the argument of authentication of origin. However, additional information is given by radiocarbon (^{14}C). When a plant carries out photosynthesis, its ^{14}C activity is at a steady state with that of the present ^{14}C atmospheric activity. Upon its death, the ^{14}C activity decreases on the basis of the radioactive decay and half life of the radio nucleide ^{14}C (5720 years). Thus, in contrast with organic molecules from living plants, fossil fuels and related molecules display no ^{14}C activity because of the extreme age of their origin.

Some examples of such a research of the natural or synthetic origin of flavoring substances can be given. Natural bitter almond oil can be adulterated by synthetic benzaldehyde. This latter molecule is the main component of the essential oil from kernels of bitter almonds, cherries, cherry laurel, apricots, peaches, and plums. The distinction between synthetic and natural benzaldehyde can be performed using both measurements of ^{14}C activity and of carbon as well as hydrogen ratios.[42,43] Benzaldehyde from natural sources shows $\delta^{13}C$ values ranging between −28.8‰ and −30.4‰ as expected from shikimic acid-derived products from C3 plants. Synthetic benzaldehyde may be prepared by oxidation of toluene; the $\delta^{13}C$ ranges then between −26.3‰ and −27.3‰. It can also be obtained by hydrolysis of benzal chloride with $\delta^{13}C$ values between -26.4 and $-31.4‰$. Benzaldehyde from natural sources cannot be distinguished from benzaldehyde from synthetic sources on the basis of $\delta^{13}C$ values. However, a differentiation is possible on the basis of $^2H/^1H$ ratios (natural: $-100‰$ to $-150‰$; from toluene oxidation: $+753‰$ to $+802‰$; and from benzal chloride: $-68‰$ to $-11‰$). In the study by D. A. Krueger, the values of ^{14}C activity were found, as expected for natural and synthetic benzaldehyde, pointing out synthetic benzaldehyde (authenticated by its $^2H/^1H$) fraudulently enriched with ^{14}C in order to simulate a natural product. The same kinds of studies have been carried out on cinnamic aldehyde and *cassia* essential oil.[44] Other important active molecules such as ethyl butyrate (orange, apple, strawberry) can be isotopically analyzed using the same techniques.[45] It is obvious that the current trend toward natural foods and beverages and the necessity to differentiate natural from synthetic flavoring substances, required by regulation and legislation, have increased the need and stimulated the development of appropriate analytical methods. Investigations of chiral compounds with regard to the determination of enantiomeric distribution have gained importance in the flavor industry. Chirality and enantiomeric ratios can be applied as criteria to differentiate between natural and nature-identical origins. It has been shown that in various natural compounds, one enantiomeric form predominates and generally synthetic organic compounds are produced as racemic mixtures. At present time, several flavoring molecules can be obtained from biotechnological processes, such as γ decalactone from *Fusarium poae* or (E)-α-ionone, and when produced by micro-organisms, they can be biosynthesized as racemic mixtures. As far as these "natural compounds" are added to natural extracts in order to reinforce flavoring, the detection of racemic forms cannot be assessed as a fraudulent addition of "synthetic" molecules. So chirospecific analysis must be performed with multidimensional capillary gas chromatography and on-line IRMS in order to obtain both enantiomeric distribution and isotope ratios. Demonstrative results have been obtained with γ decalactone.[46-49] Hener et al. have also shown for linalool analysis that while chirospecific analysis detects the R (80%):S (20%) enantiomeric ratio, indicating a blend with synthetic racemate, the amount of synthetic racemate cannot be calculated due to the possibility of partial racemization of linalool during hydrodistillation.[50] By means of enantio-IRMS analysis, the blend of linalool from different origins is unambiguously proved [R($-26.1‰$), S ($-30.6‰$)], whereas simple IRMS analysis [R + S ($-27.0‰$)] is completely unsuitable to solve the problem. These examples show that the last improvements in separate techniques (multidimensional gas chromatography, chiral analysis), coupled with the information from isotope ratios, constitute a modern and very powerful tool to differentiate natural from nature-identical flavoring compounds and to assess the origin of such widely commercialized substances.

D. Research of the Natural or Synthetic Origin of Biomolecules Fraudulently Used for Doping in Sport

With the increase of athletes' performances, new sophisticated means for doping appear for high level sport competitions. Some of them are based on the use of biomolecules which act on the metabolism and on physical performances. Among them are hormones such as testosterone and peptides like hematopoietin. Androgens are used for their anabolic actions. Consequently, testosterone is used by athletes in the hope that muscle development will be enhanced and thereafter athletic performances. Present methods used to detect the presence of exogenous testosterone in the urine of sportsmen rely upon the quantitative determination of test-osterone and epitestosterone and on the calculation of their ratio. While abnormal high ratios in that test are good indicators of administration of testosterone, this determination, taken as the only criterion, may exclude a number of suspicious cases. So, attempts have been performed to de-velop isotopic methods in order to distinguish between endogenous and exogenous testosterone on the basis of its ^{13}C content which may vary according to testosterone origin. Measurements of $\delta^{13}C$ values of testoster-one extracted from the urine of a male subject before and after receiving a dose of synthetic testosterone heptanoate were performed by GC-C-IRMS.[51] A significant difference in $\delta^{13}C$ values was observed between the pre-administration samples (-26.58 ± 0.25‰), those after administration of the drug (-30.3 ± 0.42‰), and those of the drug itself (-29.63 ± 0.52‰). These results are promising, and analytical as well as methodological improvements are presently being developed. One of the interests of the sensitivity of on-line GC-C-IRMS relies on the possibility, from a urine extract, to measure the carbon isotope ratio, not only of testosterone, but also of molecules which are either metabolic precursors of testosterone or catabolites. This isotopic determination could afford information on the isotopic coherence all along the metabolic pathway. Introduction of test-osterone with a modified isotope content would induce an isotopic dis-continuance along the metabolic route between precursors and metabo-lites of testosterone. As for other problems of authentication of origin, GC-C-IRMS and determination of isotope ratios have to be developed and improved in the important areas of sport and doping.

E. Protection of Patents, Origin of Industrial Products, Personal Isotopic Signature

If the differentiation of various origins is possible between natural and synthetic molecules or inside the vegetal kingdom, it can also be possible between molecules from various synthetic routes. So, IRMS can be an excellent tool in order to obtain the necessary data to protect patents, to prove the origin of industrial compounds, and to perform a personal isotopic signature of a commercial product.

As early as 1975, P. Bommer from Hofmann Laroche published interesting results on the use of IRMS to differentiate between various batches of Diazepam synthesized either by Hofmann Laroche Switzerland or by Hofmann Laroche U.S.[52] Isotope ratios of carbon and hydrogen were measured from various batches of Diazepam synthesized either in Switzerland or in the U.S. according to the same general process. The $\delta^{13}C$ values of the batches from Switzerland synthesized between 1967 and 1975 ranged in an interval of 2.1‰ around the mean value of about −35.5‰; the batches from the U.S. ranged between 1.3‰ around the mean value of −30.5‰. The two populations of isotope ratios values did not overlap. Thus, it was possible to easily differentiate the industrial site of production of the various Diazepam batches. Combining hydrogen isotope ratios with carbon isotope ratios makes the differentiation easier. These differences in isotopic composition mainly depend on:

- Isotopic composition of reagents, solvents, and precursors
- Kinetic isotope effects occurring during the synthetic process
- Isotopic fractionation due to physical parameters and occurring during the process (distillation, solvent evaporation, extraction, etc.)

Slight but constant modifications in a well-defined process can record inside the molecule the signature of its synthesis. So, one can admit that different methods of organic synthesis using different sources of chemicals as well as different processes, in order to obtain the same final product, would engrave a typical signature. This signature will be easily detectable by isotope ratio mass spectrometric analysis.

Such an application can be used to obtain formal information on fraudulent copies in order to protect a patent or property rights. It is clear that in order to improve the power of such a signature, it is possible to voluntarily enrich a synthetic product at low and well-defined levels in order to give a personal isotopic signature which can be read only by IRMS. Consequently, isotope ratios are the major and objective proofs of property for the owner of the substance.

V. CONCLUSION

The history of the origin of organic molecules is progressively engraved inside the molecule itself by successive and slight isotope effects, which give a characteristic isotopic composition considered as the isotopic signature of its origin. One can consider that, as far as this signature could be detected and measured with precision, accuracy and sensitivity, the legal use of this information would be possible. So, the determination of isotope ratios from small quantities of substances, the determination of isotope profiles from complex mixtures, and the coupling of enantiomeric and isotopic compositions are new and powerful tools for forensic analysis. Much work has to be performed to develop and improve the on-line analysis of isotopes other than carbon (hydrogen, oxygen, nitrogen) in order to obtain forensic information of the best quality.

ACKNOWLEDGMENTS

The author greatly thanks Mireille Buisson and Sophie Dumont for their skillful help.

REFERENCES

1. O'Leary, M. H., Madhavan, S., and Paneth, P., *Plant Cells Environ.*, 15, 1099, 1992.
2. O'Leary, M. H., *Phytochemistry*, 20, 553, 1981.
3. Henderson, S. S., Von Caemmerer, S., and Farqhuar, G. D., *Aust. J. Plant. Physiol.*, 19, 263, 1992.
4. Winter, K. and Froughton, J. H., *Flora*, 167, 1, 1978.
5. Deleens-Provent, E. and Schwebel-Dugue, N., *Plant Physiol. Biochem.*, 25, 567, 1987.
6. Monson, R. K. and Moore, B., *Plant Cell Environ.*, 12, 689, 1989.
7. Von Caemmerer, S., *Plant Cell Environ.*, 15, 1063, 1992.
8. Raven, J. A. and Farqhuar, G. D., *New Physiologist*, 116, 505, 1991.
9. Dansgaard, W., *Geochim. Cosmochim. Acta*, 6, 241, 1954.
10. Gat, J. R., *Water Resour. Res.*, 7, 980, 1971.
11. Yapp, C. J. and Epstein, S., *Earth Planet. Res.*, 34, 333, 1977.
12. Ziegler, H., in *Stable Isotopes in Ecological Research*, Rundel, P. W., Ehleringer, J. R., and Nagy, K. A., Eds., Springer-Verlag, Berlin, 1989, 105.
13. Bricout, J. and Koziet, J., *J. Agric. Food Chem.*, 35, 758, 1987.
14. Martin, G. J., Martin, M. L., Mabon, F., and Michon, M. J., *J. Chem. Soc.*, 616, 1982.
15. Caer, V., Trierweiler, M., Martin, G. J., and Martin, M. L., *Anal. Chem.*, 63, 2306, 1991.
16. Sullivan, J. J. and Quimby, B. D., *Anal. Chem.*, 62, 1034, 1990.
17. Deruaz, D., Bannier, A., Desage, M., and Brazier, J. L., *Anal. Lett.*, 24, 1531, 1991.
18. Craig, H., *Geochim. Cosmochim. Acta*, 12, 134, 1957.
19. Barrie, A., in *Stable Isotopes in Plants and Nutrition, Soil, Fertility and Environmental Studies*, IAEA, Vienna, 1991, 3.
20. Prosser, S. J., Brookes, S. T., Linton, A. L., and Preston, T., *Biomed. Mass Spectrom.*, 20, 724, 1991.
21. Freedman, P. A., Gullyon, E. C., and Jumeau, J., *Int. Lab.*, 2, 22, 1988.
22. Tissot, S., Normand, S., Guilluy, R., Pachiaudi, C., Beylot, M., Laville, M., Cohen, R., and Riou, J. P., *Diabetologia*, 33, 449, 1990.
23. Nitz, S., Weinreich, B., and Drawert, F., *J. High Res. Chromatogr.*, 15, 367, 1992.

24. Guilluy, R., Billon-Rey, F., and Brazier, J. L., *J. Chromatogr.*, 562, 341, 1991.
25. Guilluy, R., Billon-Rey, F., Pachiaudi, C., Jumeau, J., and Brazier, J. L., *Anal. Chim. Acta*, 259, 193, 1992.
26. Bricout, J. and Koziet, J., *Ann. Fals. Exp. Chim.*, 69, 845, 1975.
27. Bricout, J., Koziet, J., Derbesy, M., and Beccret, B., *Ann. Fals. Exp. Chim.*, 74, 691, 1981.
28. Krueger, D. A. and Krueger, H. W., *J. Agric. Food Chem.*, 31, 1265, 1983.
29. Krueger, D. A. and Krueger, H. W., *J. Agric. Food Chem.*, 33, 323, 1985.
30. White, J. and Doner, L., *J. Assoc. Off. Anal. Chem.*, 61, 746, 1978.
31. Doner, L., Stable isotope ratio analysis for authentication of honey, maple syrup and fruit juice in the United States, European Seminar on Authentication and Quality Assessment of Food Products, Martin, G., Ed., Nantes, 1991.
32. Krueger, H. W. and Reesman, R. H., *Mass Spectrom. Rev.*, 1, 205, 1982.
33. Bricout, J., in *Stable Isotopes*, Schmidt, H. L., Fortel, H., and Heinzinger, K., Eds., Elsevier, Amsterdam, 1982, 483.
34. Lee, H. S. and Wrolstad, R. E., *J. Assoc. Off. Anal. Chem.*, 71, 789, 1988.
35. Martin, G. J., Martin, M. L., and Mabon, F., *J. Am. Chem. Soc.*, 104, 2656, 1982.
36. Martin, G. J., Zhang, B. L., Martin, M. L., and Dupuy, P., *Biophys. Res. Commun.*, 111, 890, 1983.
37. Martin, G. J. and Martin, M. L., *Tetrahedron Lett.*, 22, 3525, 1981.
38. Martin, G. J., Guillou, C., Naulet, N., Brun, S., Cabanis, J. C., Cabanis, M. T., and Sudraud, P., *Sci. Aliment.*, 6, 385, 1986.
39. Dunbar, J. and Wilson, A. T., *Anal. Chem.*, 54, 590, 1982.
40. Liu, J., Lin, W. F., Fitzgerald, M. P., Saxena, S. C., and Shied, Y. N., *J. Forensic Sci.*, 24, 814, 1979.
41. Desage, M., Guilluy, R., Brazier, J. L., Chaudron, H., Girard, J., Cherpin, H., and Jumeau, J., *Anal. Chim. Acta*, 247, 249, 1991.
42. Butzenlechner, M., Rossemann, A., and Schmidt, H. L., *J. Agric. Food Chem.*, 37, 410, 1989.
43. Krueger, D. A., *J. Assoc. Off. Anal. Chem.*, 70, 175, 1987.
44. Culp, C. A. and Noakes, J. E., *J. Agric. Food Chem.*, 38, 1249, 1990.
45. Byrne, B., Wengenroth, V. J., and Krueger, D., *J. Agric. Food Chem.*, 34, 736, 1986.
46. Weinreich, B. and Nitz, S., *Chem. Mokrobiol. Technol. Lebensm.*, 14, 117, 1992.
47. Nitz, S., Kollmannsberger, H., Weinreich, B., and Drawert, F., *J. Chromatogr.*, 557, 187, 1991.
48. Bernreuther, A., Koziet, J., Brunerie, P., Krammer, G., Christoph, N., and Schreier, P., *Z. Lebensm. Unters. Forsch.*, 191, 299, 1991.
49. Mosandl, A., Hener, U., Schmarr, H. G., and Rautenschlein, M., *J. High Res. Chromatogr.*, 13, 258, 1990.
50. Hener, A., Braunsdorf, R., Kreis, P., Dietrich, A., Maas, E., Euler, E., Schlag, B., and Mosandl, A., *Chem. Mikrobiol. Technol. Lebensm.*, 14, 129, 1992.
51. Southan, G., Mallet, A., Jumeau, J., Craig, S., Poojara, N., Mitchell, D., Wheeler, M., and Brooks, R., Misuse of testosterone in sport: an approach to detection by measurement of isotopic abundance using GC-IRMS, 2nd International Symposium on Applied Mass Spectrometry in the Health Sciences, Barcelona, 1990.
52. Bommer, P., Moser, H., Stichler, W., Trimborn, P., and Vetter, W., *Z. Naturforsch.*, 31, 112, 1976.

INDEX